CROP PRODUCTION SCIENCE IN HORTI series

Series Editors: Jeff Atherton, Senior Lecturer in Horticulture, University of Nottingham, and Alun Rees, Horticultural Consultant and Editor, *Journal of Horticultural Science*.

This series examines economically important horticultural crops selected from the major production systems in temperate, subtropical and tropical climatic areas. Systems represented range from open field and plantation sites to protected plastic and glass houses, growing rooms and laboratories. Emphasis is placed on the scientific principles underlying crop production practices rather than on providing empirical recipes for uncritical acceptance. Scientific understanding provides the key to both reasoned choice of practice and the solution of future problems.

Students and staff at universities and colleges throughout the world involved in courses in horticulture, as well as in agriculture, plant science, food science and applied biology at degree, diploma or certificate level will welcome this series as a succinct and readable source of information. The books will also be invaluable to progressive growers, advisers and end-product users requiring an authoritative, but brief, scientific introduction to particular crops or systems. Keen gardeners wishing to understand the scientific basis of recommended practices will also find the series very useful.

The authors are all internationally renowned experts with extensive experience of their subjects. Each volume follows a common format covering all aspects of production, from background physiology and breeding, to propagation and planting, through husbandry and crop protection, to harvesting, handling and storage. Selective references are included to direct the reader to further information on specific topics.

Titles Available:
1. **Ornamental Bulbs, Corms and Tubers** A.R. Rees
2. **Citrus** F.S. Davies and L.G. Albrigo
3. **Onions and Other Vegetable Alliums** J.L. Brewster
4. **Ornamental Bedding Plants** A.M. Armitage
5. **Bananas and Plantains** J.C. Robinson
6. **Cucurbits** R.W. Robinson and D.S. Decker-Walters
7. **Tropical Fruits** H. Nakasone and R.E. Paull
8. **Coffee, Cocoa and Tea** K. Willson
9. **Lettuce, Endive and Chicory** E.J. Ryder
10. **Carrots and Related Vegetable** *Umbelliferae* V.E. Rubatzky, C.F. Quiros and P.W. Simon

This book is dedicated to Royce S. Bringhurst,
my teacher, mentor and friend

Strawberries

James F. Hancock
Department of Horticulture
Michigan State University
East Lansing
Michigan
USA

CABI *Publishing*

CABI *Publishing* is a division of CAB *International*

CABI Publishing
CAB International
Wallingford
Oxon OX10 8DE
UK

CABI Publishing
10 E 40th Street,
Suite 3203
New York, NY 10016
USA

Tel: +44 (0)1491 832111
Fax: +44 (0) 1491 833508
Email: cabi@cabi.org

Tel: +1 212 481 7018
Fax: +1 212 686 7993
Email: cabi-nao@cabi.org

A catalogue record for this book is available from the British Library, London, UK.

Library of Congress Cataloging-in-Publication Data
Hancock, James F.
    Strawberries / J.F. Hancock
       p. cm. -- (Crop production science in horticulture; 11)
    Includes bibliographical references.
    ISBN 0-85199-339-7 (alk. paper)
    1. Strawberries. I. Title. II. Series.
    SB385.H34   1999
  634´.75--dc21

99-33190
CIP

ISBN 0 85199 339 7

Typeset in Britain by Solidus (Bristol) Ltd
Printed and bound in the UK at the University Press, Cambridge

CONTENTS

The colour plate section can be found between pages 230 and 231

PREFACE

It has been over 30 years since George Darrow published his landmark book *The Strawberry: History, Breeding and Physiology*. This contribution was a major achievement, as Darrow carefully reviewed virtually all the available research on strawberry, and assembled it in a highly readable fashion. Not since S.W. Fletcher's *The Strawberry in North America* was published in 1917 had such a thorough effort been undertaken to review strawberry science. In the years following Darrow's book, numerous books and chapters have been published on the strawberry, but all have concentrated on specific aspects of strawberry science or were compilations of symposium papers. This book is an attempt to update the science reviewed in Darrow's book, and expand it to include a number of considerations left undiscussed in his previous work.

Chapter 1 is an overview of the worldwide strawberry industry. It describes where strawberries are grown and the most important cultivars of today. Since Darrow's book, cultivar use has dramatically changed across the world and major industries have appeared in Italy, Korea and Spain. California has gained dominance in North America. In writing this section, I relied heavily on the international expertise of Tom Sjulin, David Simpson, Kogi Kawasaki, Ho-Jeong Jeong and Daniel Kirchbaum.

Chapter 2 addresses the taxonomy and evolution of the strawberry. Gunter Staudt has been particularly active in organizing the genus in the last 30 years using morphological markers, and the recent use of molecular markers shows high promise in delineating taxa and phylogenies. The evolution of *Fragaria* is still very much an unfolding story, but one that is beginning to clarify.

The third chapter discusses the history of the strawberry. This topic is covered much more extensively in several chapters of Darrow's book and *A History of the Strawberry* by S. Wilhelm and J.E. Sagen, but new information is included herein on the development of the Chilean strawberry, *Fragaria chiloensis* L. Arturo Lavín provided most of this information.

Chapter 4 reviews strawberry breeding and genetics, an area where tremendous progress has been made since Darrow's review. The genetics of numerous traits have been determined, and considerable advancements have been made in breeding theory as well as biotechnological approaches to strawberry improvement. Particularly exciting information has been generated on germplasm resources and the location of horticulturally useful traits. In many instances, I had to select a portion of the available literature on a topic, and I apologize to those whose important research was left unnoted.

The fifth chapter addresses strawberry structure and developmental physiology. Only limited new information has been generated on strawberry anatomy and the climatic control of morphogenesis over the past 40 years, but considerable data have been generated on the gas exchange patterns and water relations of strawberries. A number of nutritional considerations are also considered – aspects left largely undiscussed by Darrow.

Chapter 6 describes how strawberries are cultured across the world. The 'fine tuning' of cultural methods has been an intense area of investigation since Darrow's book, and this chapter is meant to be a summary of the major cultural systems employed and their component parts. Tom Sjulin, David Simpson, Kogi Kawasaki, Ho-Jeong Jeong and Kirk Larson helped immensely with this section by describing aspects of their industries which I have not seen.

Chapter 7 highlights recent work on fruiting and postharvest physiology. Darrow described the key environmental factors regulating floral development and berry growth, but in the last two decades, numerous advancements have been made in the long-term storage of strawberry fruit, and we know much more about the biochemistry of fruit ripening. Particularly exciting strides are being made toward elucidating the major genes associated with fruit ripening and flavour.

Chapter 8 deals with diseases and pests, the one area left largely unaddressed by Darrow. Here, the most common problems are highlighted with a discussion of their symptoms, biology and control methods. I have tried to review integrated crop management strategies whenever possible. This chapter is really a summary of several excellent books including: *Compendium of Strawberry Diseases* (John Maas, ed.), *Integrated Pest Management for Diseases* (Larry Strand) and *Integrated Pest Management for Strawberries in the Northeastern United States: a Manual for Growers and Scouts* (D.R. Cooley and S.C. Schoemann, eds).

This book is meant to be an overview of all the various aspects of strawberry science and culture. It is targeted toward strawberry researchers and students of horticulture, but should be of interest to all practising horticulturalists and strawberry growers who want to be updated on the science and history behind strawberry growing. Hopefully, the book provides a portal to the huge amount of information generated on the strawberry over the last 30 years.

I would be remiss if I did not thank the people around me who made this project possible. My wife, Ann, has been unfailing in her support and understanding. The training group led by the venerable Hal Prince has repeatedly restored my energy. The people in my lab, Pete Callow, Jaimie Houghton, Chris Owens, Gary Schott and Sedat Serce, have continually challenged my preconceived opinions and made me grow. Jorge Retamales reviewed early drafts of the manuscript and was very supportive. Thanks are due also to Driscoll Strawberry Associates, of Watsonville, California, for their generous contribution towards the cost of printing the colour plates.

Finally, I would like to thank several people who have provided advice or information on particular topics, as follows:

Bob Bors, University of Saskatchewan, Saskatoon, Canada – Evolution

Randy Beaudry, Michigan State University, East Lansing, Michigan, USA – Fruit development and postharvest physiology

Marlene Cameron, Michigan State University, East Lansing, Michigan, USA – Graphic art

Adam Dale, University of Guelph, Simcoe, Ontario, Canada – Ontario culture and varieties

Chad Finn, USDA-ARS, Corvallis, Oregon, USA – Pacific Northwest culture and varieties

Stan Hokanson, USDA-ARS, Beltsville, Maryland, USA – Matted row varieties

Eric Hanson, Michigan State University, East Lansing, Michigan, USA – Nutrition and irrigation

Ho-joeng Jeong, Pusan Horticultural Experiment Station, Pusan, Korea – Korean culture and varieties

Koji Kawagishi, Hokkaido Prefectural Dohnan Agricultural Experiment Station, Hokkaido, Japan – Japanese culture and varieties

Daniel Kirchbaum, INTA EEA-Famailla, Tucuman, Argentina – South American culture and varieties

Kirk Larson, University of California, Irvine, California, USA – California culture and varieties

Arturo Lavín, Centro Experimental Cauquenes (INIA), Chile – Domestication of strawberry

Barclay Poling, North Carolina State University, Raleigh, North Carolina, USA – North Carolina and Egypt culture

Marvin Pritts, Cornell University, Ithaca, New York, USA – Matted row culture; nutrition and irrigation

Sonia Schloemann, University of Massachusetts, Amherst, Massachusetts, USA – Integrated crop management

Doug Shaw, University of California, Davis, California, USA – Californian and Mediterranean cultivars

David Simpson, Horticulture Research International, West Malling, UK – European culture and varieties

Tom Sjulin, Driscoll Strawberry Associates, Watsonville, California, USA – California and Mediterranean culture and varieties

1

THE STRAWBERRY INDUSTRY

INTRODUCTION

The cultivated strawberry, *Fragaria × ananassa*, is a regular part of the diets of millions of people. Known for its delicate flavour and rich vitamin content, strawberries are cultivated in all arable regions of the globe from the Arctic to the Tropics.

Annual world production of strawberries has grown steadily through the ages, with quantities doubling in the last 20 years to over 2.5 million metric tons (Mt; Fig. 1.1). Most of the production is located in the northern hemisphere (98%), but there are no genetic or climatic barriers preventing greater expansion into the southern hemisphere.

The USA is the leading producing nation, with approximately 20% of the world's crop, followed by Spain, Japan, Poland, Italy and the Korean Republic (Table 1.1). US production averaged nearly 750,000 metric tonnes (t) from

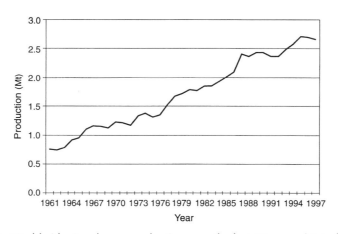

Fig. 1.1. Worldwide strawberry production over the last 35 years. (Data from FAO Production Yearbook.)

1

Table 1.1. Worldwide strawberry production (Mt × 1000). Values are means of 3-year periods. (From FAO Production Yearbook.)

Region/country	1975–1977	1985–1987	1995–1997
Africa			
Egypt	–	23.2	35.5
Morocco	0.2	0.5	10.0
South Africa	2.7	4.4	2.6
Asia			
China	–	5.3	6.5
Japan	168.6	202.1	206.5
Korean Republic	22.8	61.5	170.8
Europe			
Austria	7.5	15.4	11.9
France	71.9	93.9	78.2
Italy	146.6	175.0	179.4
Germany	52.4	84.6	75.0
Hungary	18.4	14.5	12.0
Ireland	3.3	3.8	5.0
Netherlands	19.9	21.8	25.0
Poland	165.6	271.0	184.9
Portugal	2.3	2.1	2.5
Romania	27.3	30.6	12.6
Spain	30.8	194.4	265.3
Yugoslavia (area)	14.1	30.3	37.5
UK	41.4	51.8	41.7
Oceania			
Australia	2.7	4.9	11.3
New Zealand	3.2	4.6	3.4
Scandinavia[a]			
Finland	3.8	9.5	10.8
Norway	15.9	17.7	15.5
North America			
Canada	18.4	32.1	29.1
Guatemala	–	1.1	2.7
Mexico	87.4	57.6	115.9
USA	270.9	477.1	736.1
South America	11.6	34.2	62.7
Argentina	3.9	6.0	8.3
Brazil	0.5	2.3	2.3
Chile	–	5.5	15.8
Colombia	–	4.0	15.1
Ecuador	–	8.1	0.6
Peru	3.8	3.4	12.9
Venezuela	2.6	4.8	4.0
USSR (area)	90.7	124.3	145.1
Russia	–		118.7
Ukraine	–		20.6

Table 1.1. *Continued*

Region/country	1975–1977	1985–1987	1995–1997
West-Central Asiatic			
Greece	8.8	6.2	7.3
Israel	5.0	11.6	13.0
Turkey	17.4	36.2	95.0

[a]Sweden produces about 20,000 t year^{-1} (Martinsson, 1997), but does not appear in the FAO statistics.

1995 to 1997, whereas Spain produced over 265,000 t, Japan 206,000 t, Poland 184,000 t, Italy 179,000 t and the Korean Republic 170,000 t. The industries in Spain (861%), the Korean Republic (649%) and the USA (172%) have grown steadily over the last two decades. Spain had the fastest growing industry in the 1980s, whereas Korea showed the fastest growth in the 1990s. Production levels in Japan, Italy and Poland have declined in the last decade, after dramatic increases in the 1970s and 1980s.

Overall European production stagnated at 1,193,685 t in 1995–1997, after relatively strong growth from 701,699 t in 1975–1977 to 1,097,605 t in 1985–1987 (FAO Production Statistics). In the last decade, only Spain (70%), Ireland (50%), the Netherlands (26%) and the Yugoslavian area (160%) have shown strong increases in production (Table 1.1). Production in most other European countries has dropped to mid-1970s levels after a period of relatively strong growth in the 1980s. Germany has shown a modest surge in production in the last few years. In spite of the recent production decreases in Poland, it still has by far the largest area of strawberries under cultivation with over 50,000 ha, followed by Spain and Italy with approximately 10,000 ha and the UK with over 6000 ha (Rosati, 1991). The highest strawberry yields are found in Spain where they routinely exceed 20 t ha^{-1}, with Italy not far behind. Polish yields are quite low averaging about 5 t ha^{-1}.

Strawberry production in the United States has steadily increased from 270,900 t in 1975–1977, to 477,100 t in 1985–1987 and 736,100 t in 1995–1997. Total strawberry acreage in the US has remained around 21,000 ha over the last 30 years, with the most hectarage being found in California at about 10,000 ha, followed by Oregon at about 2500 ha and Florida at approximately 2000 ha. Although acreage has remained static, average yields per acre have increased nearly threefold from nearly 8000 t ha^{-1} to over 22,000 t ha^{-1} (USDA Economic Research Service, 1995). Most of this yield increase has occurred in California and Florida, where yields can exceed 30,000 t ha^{-1}. Most of the fruit produced in North America is used domestically, although a significant proportion is shipped to Japan.

In the mid-1990s, California was by far the leading producing state in the USA, with nearly 80% of the production. Florida (10%), Oregon (5%) and Washington (2%) followed far behind (US Economic Research Service, 1995).

California's share of the market grew from 30% in 1950 to about 60% in the early 1970s to nearly 80% in the 1990s (Bringhurst *et al.*, 1990). Since the mid-1970s, Florida's share has increased from 4 to 10% and that of Oregon has fallen from 14 to 5%. Production has declined or remained stable in most other states, with the exception of North Carolina which has shown a dramatic increase in the 1990s and may now be the third highest producing state.

Canadian strawberry production almost doubled in the late 1970s and early 1980s after a collapse in the 1960s, but has held relatively stable over the last 20 years at around 30,000 t on almost 7500 ha (Table 1.1). Quebec and Ontario are the leading producing provinces with over 2500 ha in production. British Columbia is third with about 1000 ha. Yields have held relatively stable over the last 20 years.

MAJOR CULTURAL SYSTEMS

There are two major open production systems in the world – annual hills and matted rows (Plates 1 and 2). These are fully described in Chapter 6. The annual systems are planted in the summer or late autumn/early winter and feature raised beds, plastic mulch, trickle irrigation and 1–2 year production seasons. The matted row systems are planted in the spring and feature, flat beds, straw mulch, overhead irrigation and three to five production seasons. Most worldwide production is in the open, although tunnels and greenhouses are important in some areas (Table 1.2).

Because of these cultural manipulations and natural climatic variation, strawberries are produced somewhere in the world at all times. The highest quantities are generated during the late spring and early summer months of the northern hemisphere, primarily in California, Spain and southern Italy. Summer production is highest in the cooler regions of Europe and North America. Florida and Japan lead winter production, with significant quantities also being produced in Korea, Israel and California.

CHARACTERISTICS OF THE MAJOR PRODUCING REGIONS

Africa

Egypt has by far the largest strawberry industry in Africa, although acreage in Morocco has been rapidly growing. The plants in these regions are grown in high density, annual systems with plastic mulch and trickle irrigation (T. Crocker, Florida, 1997, personal communication). The climate is hot Mediterranean.

There are over 2000 small, traditional growers in Egypt that produce fruit

Table 1.2. Characteristics of the major production regions in North America and Europe.

Region	Major production systems	Primary cultivars
Africa		
Egypt	Annual in open	Chandler, Oso Grande, Selva and Sweet Charlie
Asia		
Japan	Annual in tunnels	Nyoto and Toyonoko
Korea	Annual in tunnels	Suhong, Sistakara and Akihime
Australia	Annual in open	Chandler, Pajaro, Parker and Selva
Europe		
Spain and southern Italy	Annual in tunnels	Camarosa, Pajaro, Oso Grande, Chandler and Tudla
Northern Italy and S.W. France	Annual in tunnels and open	Elsanta, Marmolada and Gariguette
Germany	Annual and matted row in open; some tunnels	Elsanta
The Netherlands and Belgium	Annual in tunnels and greenhouses	Elsanta
UK and Republic of Ireland	Annual in tunnels; matted rows in open	Elsanta, Symphony, Pegasus and Honeoye
Poland and eastern Europe	Matted rows in open	Senga Sengana
Russia and northern Europe	Matted rows in open	Senga Sengana, Gorella and Elsanta
Scandinavia	Matted rows in open; some tunnels	Senga Sengana, Korona, Honeoye, Zepyr, Bounty and Dania
North America		
California	Annual in open	Camarosa, Selva and Seascape
Florida	Annual in open	Sweet Charlie and Camarosa
Pacific NW	Matted row in open	Totem
Southern USA	Annual in open	Chandler
Lower midwestern and eastern US	Matted row in open	Earliglow, Honeoye, Kent, Annapolis, Jewel and Cavandish
Upper midwestern US and eastern Canada	Matted row in open	Kent, Glooscap, Veestar, Bounty, Annapolis and Honeoye
Mexico	Annual in open	Camarosa, Selva and Seascape
West-central Asiatic		
Turkey	Annual and matted row in open; some tunnels	Douglas, Pajaro, Cruz and Senga Sengana

for local, fresh consumption; however, a few large growers are now producing fruit for the export markets in Europe. The traditional growers set cold-stored (frigo) plants in late August to mid-September at densities of 50,000 plants ha^{-1}. The export growers set fresh plants in the autumn from 20 September to 10 October, at much higher densities of 100,000 plants ha^{-1}. The frigo plants are harvested from mid-February to June, whereas the fresh plants are harvested from December through May.

Asia and Japan

The strawberry is the major fruit crop in Japan (Oda, 1991), and is growing in popularity in Southeast Asia, particularly in Korea. Acreage in Japan has gradually dropped over the last decade, but productivity has held stable. Strawberries are grown all over Japan, with the highest concentrations in Kanto, Chubu and Kyushu (Fig. 1.2). The climate of the major production

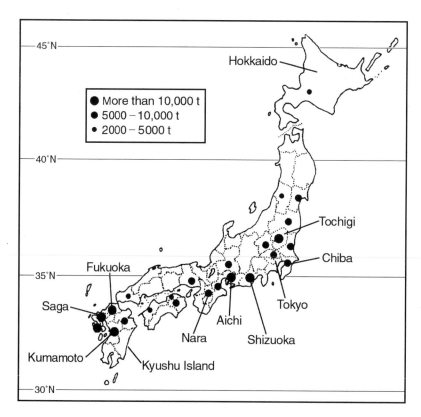

Fig. 1.2. Major production areas in Japan (adapted from Mochizuki, 1995).

regions in Japan are mild; temperatures remain above freezing during the coldest months and can often exceed 30°C. Annual rainfall approaches 2000 mm, with most falling in the summer from June to December.

Most of the production in Japan comes from intense annual, hill systems under high, plastic tunnels. In some places, CO_2 enrichment is employed to enhance productivity (Oda, 1991). Intense methods of forcing are utilized involving transplantation, gibberellic acid application and the regulation of temperature and photoperiod (Ito and Saito, 1962; Oda, 1991). Most plants are grown in the ground, although some plants are grown hydroponically and a unique system of culture utilizing cement blocks has been developed in the Shizuoka Prefecture (Oda and Kawata, 1993). Fruit is produced primarily for the wholesale market, starting in November, with peak production between January and April. The latest fruit are produced from late April to early July in the northern island of Hokkaido in unheated plastic tunnels (K. Kawagishi, Japan, 1998, personal communication). Off-season fruit are now primarily imported from California, New Zealand, Taiwan and South Korea.

In Korea, most strawberries are grown below latitude 37° where winter production is possible under protection without supplemental heating. The climate in these regions is very similar to Japan. In the early 1980s, most acreage was planted in open fields, but now most is under protection, with 1 m high plastic tunnels predominating (Ho-joeng Jeong, Korea, 1998, personal communication). Plants are generally planted in the early autumn in annual systems for winter and spring production. Most farmers produce their own runner plants, often collecting them from overwintered mother plants. Most of the fruit is used locally, although a small amount of processed fruit is shipped to Japan.

China

Strawberry culture in China is in its early stages, but is rapidly increasing (Wang and Tang, 1994). Currently about 2000 ha are planted with a total annual yield of 20,000 t. The primary growing areas are in Hebei, Liaoning, Shandong, Jilin, Heilongjiang, Zhajiang and Hubei provinces. Most farms are very small (0.07–0.2 ha) and all field management is done by hand. Most production is in the open, with increasing use of low polyethylene tunnels. Plants are produced primarily in close spaced, annual systems on raised beds covered with black plastic. Fresh dug daughter plants are set in the autumn. Strawberries are harvested from late April to mid-June, mostly for the fresh market (80%).

Europe and Russia

Nine major production regions can be identified in Europe: (i) Spain, southern France and southern Italy; (ii) northern Italy and southeastern France;

(iii) Belgium and the Netherlands; (iv) Germany; (v) the UK and Ireland; (vi) Poland; (vii) eastern Europe; (viii) Russia and northern Europe (the former USSR), and (ix) the Scandinavian countries (Hancock and Simpson, 1995).

Spain and southern Italy have continental climates and mild Mediterranean winters. More than half the strawberry acreage in Spain is in the southwest (West Andalucia, Huelva), followed by the Mediterranean coast (Valenciana Community, East Andalucia and Cataluna). The majority of the hectarage is under small tunnels for early production of fresh market fruit. Plants are grown in intense annual systems on raised beds with preplant soil fumigation, drip irrigation and high planting density (López-Aranda and Bartual, 1999). The strawberries in Huelva are set primarily in the late autumn as fresh plants. Ripening starts in early February and ends in July, with limited quantities of fruit produced until December. April is the peak month of production. France, Germany and the UK are the leading importers of Spanish berries.

Northern Italy and France have relatively mild temperatures, but occasional cold winters. Significant harvests begin in April under tunnels and end in the open about July, with limited production all summer. Over 60% of the

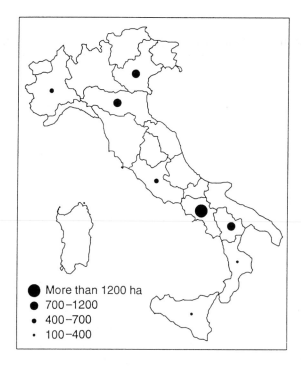

Fig. 1.3. Major production areas in Italy.

production in France is in the southwest, and most of it originates from annual systems under low tunnels (Rosati, 1991). About 40% of the Italian production is in the north, on approximately half the hectarage of the south (Fig. 1.3) (Faedi *et al.*, 1997). Annual systems dominate in Italy, with over 60% of the acreage under primarily walk-in tunnels. Campania is the major producing region, where almost all the hectarage is protected. Peat-bag culture is growing in popularity in the northern mountain areas of Italy (Piemonte and Trentino regions).

The Netherlands and Belgium have cold continental climates that are modulated by the ocean. Here they produce fruit virtually year-round. Approximately 10% of Dutch strawberries and 20% of Belgian strawberries are produced in greenhouses and plastic houses. High and low tunnels are used to extend the outdoor season. Plants in greenhouses are maintained at moderately high densities, grown in containers and trickle-irrigated. Harvest dates are made almost continuous by staggering planting dates both outside and under protection, and by altering the temperature in tunnels and greenhouses.

The UK and the Republic of Ireland have mild, maritime climates. They use a wide array of protected systems to maximize the harvest period. Floating mulches are commonly placed over fields in the late autumn to hasten the fruiting season of short-day cultivars. Likewise, low tunnels ('polythene cloches') and, more often, portable high tunnels (called 'French tunnels' in the UK) are put over short-day plants in February to advance their crops, or they are placed over day neutrals in August to extend the season into November. Many growers plant in May or June, take a 60-day crop the same year, and then harvest the plants for one or two seasons. Glasshouses are utilized for late autumn production. Matted rows in the open are still popular on pick-your-own (PYO) farms, but runnerless culture under protection now dominates commercial production. Virtually all fruit goes to the fresh market, with the fastest growing sector being supermarkets.

Germany has a cold continental climate. Most of the production is in open fields, but tunnels are frequently used to advance the harvest season. Open grown, annual hill systems predominate in the milder, southern regions, and matted rows in the cooler, northern areas.

Poland and eastern Europe have continental climates with moderate to severe winters. Strawberry production throughout Poland is relatively scattered, but major regions can be identified in the northern, southeastern and southwestern parts of the country . The highest amount of hectarage is planted in the state of Klimontow in the southeastern portion of the country. Nearly 1000 farmers grow strawberries in this region, with farms ranging in size from 0.02 to 0.50 ha. Significant hectarages of strawberries are also found in Hungary, the northern Balkan states and Romania (FAO Production statistics; Stanisavljević *et al.*, 1997). Almost all the strawberries grown in these regions are in open, matted row systems and yields average below 5 t ha^{-1}.

Fruit is harvested in May in the area of the former Yugoslavia and June in Poland, and is exported primarily for processing to Europe and to a limited extent to the USA.

Scandinavia has a continental climate with severe winters. Fruit are produced primarily for the fresh market in late June to early July, but in northern areas and at high elevation, fruit is produced in late September (Nes, 1997). Most of the acreage is in open, matted rows, although polyethylene plastic soil covers are used for early production on about 30% of the hectarage in Sweden (Martinsson, 1997).

Russia and northern Europe have a continental climate with severe winters. Fruit are generally produced on matted rows with very low inputs. Some of the major production regions are near Moscow and Leningrad, North Caucus, Ukraine and Middle Asia (Govorova, 1992). Significant acreages also exist in the Ukraine and Latvia (Bite *et al.*, 1997).

North America

Most of the North American production can be separated into six geographical regions based on the cultivars grown (Hancock and Luby, 1995). These are: (i) California; (ii) Florida; (iii) southern US; (iv) Pacific Northwest; (v) lower Midwest and eastern US; (vi) upper Midwest and eastern Canada. Virtually all the production in North America is in the open, with only a limited amount of protected culture employed.

About two-thirds of the acreage in California is planted in the central coast and Santa Maria Valley (Fig. 1.4). Most of the rest of the California acreage is found in the Oxnard Plain, with other important commercial acreage being located in the South Coast and Central Valley. The general climate of these regions is Mediterranean with mild winters and dry summers. Southern California production is dominated by short-day cultivars that ripen fruit from early January until April or May. The central coast begins production in April with day-neutral types and continues until October or November. Runnerless culture is used on raised beds covered with mostly clear plastic. Dormant plants are set in the late autumn or winter depending on location and purpose. Most of the fruit is sold for the fresh market, but California is still the leading producer of processed fruit as a clean-up operation through sheer volume.

Most of the production in Florida centres around the Plant City area, east of Tampa. The plants are grown without runners, on raised beds in black plastic mulch. Usually, fresh dug plants are set in the early winter. The fruit is used primarily for fresh shipment during the late winter months.

The southern US is distinguished by ample rainfall, highly fluctuating winter temperatures and hot, humid summers. The major production state is North Carolina, with Louisiana a distant second. Matted row systems are still

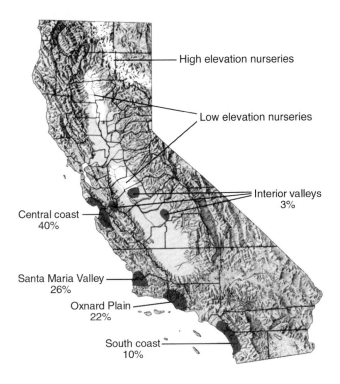

High elevation nurseries

Low elevation nurseries

Interior valleys
3%

Central coast
40%

Santa Maria Valley
26%

Oxnard Plain
22%

South coast
10%

Fig. 1.4. Major production areas in California. (Reprinted by permission from Strand, L.L. (1994) *Integrated Pest Management for Strawberries,* Publication 3351, University of California Statewide Integrated Pest Management Project, Davis.)

common, but close-spaced runnerless systems are becoming dominant (Hokanson and Finn, 1999). In North Carolina, plasticulture acreage has increased from about 2 ha in 1983 to nearly 500 ha in 1997 (Poling, North Carolina, 1998, personal communication). Most of the fruit grown in the south is sold in local markets, as PYO and as ready picked.

The Pacific Northwest can be characterized as having mild, cool maritime climates with ample moisture. Most strawberry production is in Oregon, but Washington and British Columbia also produce significant quantities of fruit. Most of the strawberries are grown on spring-set matted rows, although some runnerless culture is employed. The bulk of the fruit is processed, both locally and shipped to the east and Midwest.

The climate of the lower midwestern and eastern parts of the US is a continental one with cold winters, relatively hot summers and temperature extremes. New York, Michigan and Pennsylvania are the leading producing states. Rainfall is sporadic and at times heavy. Spring planted matted row systems of culture predominate in this region on flat beds set in the spring.

Most of the production is in the early summer months, with most fruit being sold locally through farm markets or PYO, although a significant quantity of the fruit in Michigan is shipped to the fresh market and processed.

The upper Midwest and eastern Canada is characterized by cold winters and very short, cool summers. Quebec and Ontario are the leading producers, followed by Nova Scotia, Wisconsin and New Brunswick. The winter low temperatures and summer high temperatures in Nova Scotia and New Brunswick are much more moderate than those of the upper Midwest (−30 vs. −40°C; 32 vs. 38°C); however, both these locations have very short frost-free periods (120–140 days) and late floral induction periods of < 14 h of day length (18 August). In these regions, matted rows are set in the spring and held for 3 years. Fruit is produced in the early summer for primarily the fresh market, but efforts are being made to further exploit processed outlets (Khanizadeh *et al.*, 1995).

The strawberry industry of Latin America is centred in the Michoacan and Gunanajuato states, northwest of Mexico City. Significant acreage also exists in Guatemala and Costa Rica. Runnerless culture is used on raised beds using predominantly California cultivars. Peak production comes in March in Mexico, although Guatemala and Costa Rica also produce fruit in the winter months. Most of the fruit is frozen and exported to the USA for processing.

West-central Asiatic

The most substantial industries in this vast area are in Israel and Turkey. In Israel, intense summer planted, annual systems predominate with an emphasis on growing strawberry types that result in earlier autumn fruit production and harvest (Izhar, 1997). In Turkey, traditional matted row systems are still used in the largest production region of Marmara with most of the fruit being processed into jam and marmalade. However, annual systems have become important in the Mediterranean and Aegean regions. Plants are set in the winter as freshly dug runners rooted in the nursery under mist, or in the summer as either frigo plants or pot-rooted runners (Kaşka, 1997). Most of the production is on raised beds on black plastic mulch in unheated, high tunnels (Kaşka *et al.*, 1997). The first harvests begin in the coastal areas in late February or March, with a peak in April or May. The largest berries are sold fresh to local markets, and the smaller ones are processed. Virtually all farms in Turkey are smaller than 0.1 ha.

Southern hemisphere

Far less strawberry fruit is produced in the southern than northern hemisphere, but significant pockets of production do exist in Australia, Colombia,

Peru, Argentina and Chile. Annual plasticulture systems are generally employed, with fresh plants being planted in the autumn and frigo plants in the summer (Kirschbaum, Argentina, 1997, personal communication). The climates in the major production regions vary from subtropical in Colombia, Peru and Argentina to both mild, temperate and hot Mediterranean in parts of Chile. About half the fruit is consumed fresh locally and most of the rest is processed. A little fresh fruit is shipped to North America, although most of that which travels north is processed. Ecuador is a relatively minor producer on only about 350 ha, but is unique in that its high elevation and constant 12 h photoperiod has resulted in year round production on short-day plants, with peaks every 3 months (Finn *et al.*, 1998).

In Australia, most of the fruit is produced in open, annual, plasticulture systems using fresh plants (Zorin *et al.*, 1997). Strawberries are grown at numerous locations across the country, and as a result, fruit are produced almost year-round. California short-day and day-neutral cultivars predominate, although native bred types are beginning to appear. There is also a significant hydroponic industry in Australia where strawberries are grown in the open on sloping, 1.2 m high tables covered with corrugated asbestos sheets with channels for nutrient flow. Almost all the strawberry fruit produced in Australia is used domestically.

VARIETIES

The most widely planted cultivar in the world is 'Camarosa', released from the University of California programme (Table 1.3). It is important in all climates with mild winters (Florida, southern USA, Australia, Italy, New Zealand, South America, South Africa, Mexico and Spain). California-bred 'Selva', 'Chandler', 'Oso Grande' and 'Pajaro' are also very important in these regions. 'Honeoye' has probably gained the broadest foothold in world climates with cold winters, followed by 'Earliglow' which dominates eastern US acreage and Dutch bred 'Elsanta' which predominates in non-Mediterranean Europe.

The cultivar situation in the warm climates of the USA has changed dramatically over the last 10–15 years. In California, the dominant public-bred cultivars of the mid-1980s 'Douglas' and 'Pajaro' were replaced, first by 'Chandler', 'Selva' and 'Oso Grande', and most recently by 'Camarosa'. 'Seascape' has also had an important impact in the 1990s. Among the privately bred cultivars, 'Balboa', 'Swede' and 'Heidi' of Driscoll Strawberry Associates were displaced by 'Commander', 'Lido', 'Key Largo`, 'E26', 'Coronado` and 'San Miguel'. Other privately developed varieties that are currently important in California are 'Catalina' (introduced by New West Foods, Inc.), PSI 118 and PS 592 (the later two from Plant Sciences, Inc.). The cultivar situation in California will continue to rapidly change as several new releases show promise including 'Diamante', 'Aromas', 'Gaviota' and some

Table 1.3. The most popular strawberry cultivars in the last 25 years.

Addie	Northern Italy (Centro Attivia Vivaistiche, Italy)
	Regionally popular; early; large; only moderate firmness; high yielding
Aiko	California (University of California)
	Currently little grown; particularly long fruiting season; small, malformed fruit during the early part of season.
Allstar	Eastern, mid-Atlantic and midwestern US (USDA-Maryland)
	Currently widely grown; mid-season; large, light coloured, symmetrical fruit; resistant to red stele, leaf scorch and powdery mildew
Annapolis	Upper midwestern US and eastern Canada (Agriculture Canada Research Station, Nova Scotia)
	Currently widely grown; early; large to medium, light coloured fruit; resistant to verticillium wilt
Apollo	Southern US (USDA-Maryland and North Carolina Agriculture Experiment Station)
	Popularity declining; late; symmetrical, deep scarlet fruit; medium vigour; multiple leaf disease resistances
Aromas	California (University of California)
	Currently widely grown; day-neutral; large, firm conic fruit; very glossy; high yielding; resistant to *Phytophthora cactorum*, Anthracnose crown rot and powdery mildew; tolerant to two-spotted spider mite
Atlas	Southern USA (USDA-Maryland and North Carolina AES)
	Popularity declining; midseason; glossy, medium red fruit; good freezing quality; vigorous; resistant to red stele
Avila	California (Driscoll Assoc., California)
	Regionally popular; very good size and appearance; orange-red to light-red colour; tart; rain resistant
Balboa	California (Driscoll Assoc., California)
	Popularity declining; early with extended season, symmetrical, flavourful fruit with medium-red interior; softens in storage
Belrubi	Italy / France (INRA-Montfavet, France)
	Popularity declining; late mid-season; large primaries but size drops; outstanding fruit quality; high vigour; adapted to high pH soils
Benton	Pacific Northwest (USDA and AES-Oregon, Washington AES)
	Regionally popular; late mid-season; large with light internal colour and poor texture; high yields; resistant to red stele
Blomidon	Upper midwestern US and eastern Canada (Agriculture Canada Research Station, Nova Scotia)
	Popularity declining due to June yellows; late mid-season; difficult to cap
Bogota	UK (Institute for Horticulture and Plant Breeding, the Netherlands)
	Popularity declining; late fruiting; large with orange–red colour; productive; fruit size drops in second year; June yellows

Table 1.3. *Continued*

Bolero	UK (Horticulture Research International, UK)
	New release; day-neutral; medium size; glossy, orange–red colour; excellent fruit quality and shelf life; moderately resistant to crown rot and verticillium wilt
Bounty	Eastern Canada/Scandinavia (Agriculture Canada Research Station, Nova Scotia)
	Regionally popular; late; medium sized fruit; vigorous and productive; susceptible to mildew and verticillium wilt
Calypso	UK (Horticultural Research Institute, UK)
	Popularity declining; day-neutral; irregular fruit; abundant runners
Camarosa	California (University of California)
	Most important cultivar in the world; early; very large, firm fruit; productive; vigorous
Captiva	California (Driscoll Associates)
	New release; early; good appearance; ships well
Cardinal	Southern US (Arkansas AES)
	Popularity declining; mid-season; large with fair flavour; concentrated ripening; caps easily; good for processing; multiple leaf disease resistances
Catalina	California (New West Fruit Corporation)
	Early; high yields; well shaped, unblemished fruit; consistent fruit size; steady producer; tolerant to verticillium wilt, *Phytophthora* spp. and non-fumigated conditions
Cavendish	Eastern Canada (Agriculture Canada Research Station, Nova Scotia)
	Popularity declining; mid-season; large, flavourful fruit; deep red, but uneven berry colour; resistant to red stele and verticillium wilt
Chamby	Eastern Canada (Agriculture Canada Research Station and Macdonald College, Quebec)
	New release; medium-sized, dark-red fruit with white shoulders; resistant to several leaf diseases and red stele; very winter hardy
Chandler	California/all mild climates (University of California)
	Major worldwide cultivar; large, firm fruit with occasional white shoulders; high yields; virus tolerant
Clea	Southern Italy (Consorzio Italiano Vivaisti, Italy)
	New release; early, large and firm; orange–red colour; fair flavour
Commander	California (Driscoll Associates, California)
	Regionally popular; large symmetrical fruit with light colour; excellent flavour; long season; good shelf life; resistant to crown rot
Coronado	California (Driscoll Associates, California)
	Regionally popular; very good size and appearance; orange–red to light-red colour; tart; rain resistant
Dabreak	Southern US (Louisiana)
	Currently little planted; early; large and attractive fruit; medium red; resistant to leaf spot

Table 1.3. *Continued*

Dana	Northern Italy (Centro Attivia Vivaistiche, Italy)
	Popularity declining; late mid-season; very large fruit; only moderate firmness; resistant to verticillium wilt
DarSelect	Central France (CIREF, France)
	Regionally popular; mid-season; productive; large and firm; mid-red colour; very sweet
Delmarvel	Midwestern, mid-Atlantic and eastern US (USDA-Maryland)
	New release; early mid-season; large and symmetrical; excellent aroma; resistant to multiple leaf diseases and red stele
Diamante	California (University of California)
	Currently widely grown; day-neutral; large, glossy and firm fruit; long rounded conic; light-coloured; high yielding; resistant to powdery mildew; tolerant to two-spotted spider mite
Domanil	Belgium (Station d'Amelioration des Plantes Fruitières et Maraichères, Belgium)
	Currently little planted; late season; productive; large fruited and distinctly aromatic; moderate firmness
Douglas	California and Mediterranean climates (University of California)
	Currently little planted; early; large fruited and high yielding; moderately soft and seedy
Dover	Florida (Florida AES)
	Popularity declining; early; resistant to *Colletotrichum* and anthracnose crown rot
E 26	California (Driscoll Associates, California)
	Regionally popular; everbearer; uneven colour; only moderate firmness; consistent shape; medium, variable size
Earliglow	Eastern, mid-Atlantic and midwestern US (USDA-Maryland)
	Currently widely grown; early; excellent flavour; good for processing; moderate productivity and size; resistant to multiple leaf and root diseases and botrytis fruit rot
Elsanta	Cool regions of Europe (Institute for Horticultural Breeding, the Netherlands)
	Currently widely grown; late; large, firm berries with orange–red colour
Elvira	Germany/the Netherlands (Institute for Horticultural Breeding, The Netherlands)
	Regionally popular; early; large and soft, with orange–red colour; resistant to crown rot
Eris	Southern Italy (Consorzio Italiano Vivaisti, Italy)
	New release; early; glossy light red exterior and orange–red interior; excellent flavour and intense aroma
Eros	UK, France (Horticultural Research International, England)
	New release; midseason; large, firm fruit; glossy with medium red internal colour; resistant to red stele; susceptible to crown rot
Florence	UK (Horticultural Research International, England)
	New release; late season; productive with large, firm, deep-red

Table 1.3. *Continued*

	fruit; sweet flavour; vigorous; resistant to *Verticillium*, powdery mildew and crown rot
Gariquette	Central France (INAA-Montfavet, France)
	Regionally popular under plastic tunnels ; beautiful appearance; mid-red and moderately firm; excellent flavour
Gaviota	California (University of California)
	Currently widely grown; large, firm fruit; rounded, conic; moderately dark; good flavour; resistant to powdery mildew and anthracnose crown rot; tolerant to two-spotted spider mite
Glooscap	Midwestern US/eastern Canada (Agriculture Canada Research Station, Nova Scotia)
	Currently widely grown; large, dark red; average firmness
Gorella	Northern Italy to Denmark (Institute for Horticultural Breeding, the Netherlands)
	Popularity declining; early; large, red and glossy; low yields
Guardian	Midwestern US (USDA-Maryland)
	Popularity declining; midseason; large, seedy fruit with green tips; resistant to leaf scorch, mildew, red stele and verticillium wilt
Heidi	California (Driscoll Associates, California)
	Currently little planted; symmetrical, soft fruit with light colour; long shelf life; resistant to anthracnose
Hokowase	Northern Japan (Hyogo AES)
	Regionally popular; high yielding; good flavour; a little soft; very susceptible to *Verticillium* and anthracnose; moderately resistant to powdery mildew
Honeoye	Midwestern and eastern US/eastern Canada/UK (New York AES)
	Currently widely grown; early; high yielding; large and dark with only fair flavour; good for processing
Hood	Pacific Northwest (USDA and AES-Oregon)
	Regionally popular; midseason; excellent flavour and internal colour; excellent for processing; caps easy; multiple resistances to leaf and root fungal diseases
Idea	Northern Italy (Centro Attivia Vivaistiche, Cesena, Italy)
	Regionally popular; late mid-season; high yield; very large, orange–red fruit; moderately firm; high vigour; resistant to soil-borne fungi and *Colletotrichum acutatum*
Jewel	Lower midwestern and eastern US (New York AES)
	Widely grown; late midseason; large and attractive; glossy, bright red; flavourful; drought tolerant
Joe Reiter	California (Driscoll Associates, California)
	Currently little planted; large, symmetrical fruit; high flavour; light colour; only moderately firm
Kent	Midwestern and eastern US/eastern Canada (Agricultural Canada Research Station, Nova Scotia)
	Currently widely grown; mid-season; medium red throughout, but only moderate firmness; high yields; very winter hardy

Table 1.3. *Continued*

Key Largo	Florida, California (Driscoll Associates, California) Currently widely grown; large to medium sized, symmetrical, conic fruit; light internal colour with white core; rain resistant; ships very well
Korona	The Netherlands/Scandinavia (Institute for Horticultural Plant Breeding, the Netherlands) Regionally popular; midseason; large and dark red; modest firmness and shelf life
Lateglow	Midwestern and eastern US (USDA-Maryland) Regionally popular; late; large with medium soft fruit; deep glossy scarlet; juicy, aromatic; tolerant to grey mould, leaf spot, leaf scorch and leather rot; resistant to red stele and verticillium wilt
Lido	California (Driscoll Associates, California) Regionally popular; good size and flavour; sweet with high aromatics; modest productivity and size
Marmolada	Northern Italy (Consorzio Italiano Vivaisti, Italy) Regionally popular; late mid-season with occasional second crop; large and red throughout; sweet; excellent shelf life
Mesabi	Upper Midwest (USDA-Maryland and Minnesota AES) New release; mid-season; large fruited with outstanding fruit quality; winter hardy; vigorous; resistant to common leaf and soil diseases
Mira	Midwestern US and eastern Canada (Agriculture Canada Research Station, Nova Scotia) Recent release; mid-season; bright, orange–red fruit; holds colour well; fair flavour; resistant to common leaf diseases and red stele
Mirador	Florida (Driscoll Associates) Recent release; early; good appearance; good shipper
Miranda	Southern Italy (Consorzio Italiano Vivaisti Co., Italy) New release; midseason; large with glossy orange–red exterior and interior colour; tolerant to salinity and stress
Miss	Northern Italy (Centro Attivia Vivaistiche, Italy) New release; early; very large, bright red fruit
Mohawk	Northeastern US and Ontario, Canada (USDA-Maryland) New release; very early; medium size and irregular; resistant to multiple leaf diseases and red stele
Noreaster	Northeastern and mid-Atlantic US (USDA-Maryland and New Jersey AES) New release; early; large and firm; strong flavour and aroma; resistant to red stele and most leaf diseases
Nyoho	Eastern Japan and Korea (Tochigi AES – Japan) Widely grown under PVC tunnels; very attractive, glossy appearance; firm; very sweet; excellent for cake
Oso Grande	California/all mild climates (University of California) Currently widely grown worldwide; very large, firm fruit; hollow centre; high yields

Table 1.3. *Continued*

Ostara	Germany/the Netherlands (CPRO-DLO, the Netherlands)
	Regionally popular; everbearing; productive; good flavour; small and soft
Pajaro	California/all mild climates (University of California)
	Popularity declining; firm, symmetrical fruit; high yields; virus tolerant
Pandora	UK (Horticultural Research International, England)
	Popularity declining; very late; large, relatively firm fruit; pistillate; resistant to mildew
Pegasus	UK (Horticultural Research International, England)
	Productive; mid-season; large fruit, but only moderately firm; can become dark; resistant to verticillium wilt and races of red stele
PS-592	California (Plant Sciences, Inc.)
	Vigorous with high yields; large, juicy berries; excellent flavour; moderate firmness; excellent holding qualitites; long harvest season
PSI-118	California (Plant Sciences, Inc.)
	Medium to large, smooth and conic, relatively dark-coloured fruit with excellent flavour and firmness
Rapella	Germany/the Netherlands/UK (CPRO-DLO, the Netherlands)
	Popularity declining; everbearing; productive, large berries; only moderately firm; susceptible to powdery mildew
Rachel	Israel (The Volcani Centre, Israel)
	Locally important; infra-short-day type with no chilling requirement
Raritan	Midwestern US (New Jersey AES)
	Popularity declining; mid-season; only fair flavour; high yields
Redcoat	Eastern Canada (Canada Department of Agricultural Research, Ottawa, Ontario)
	Popularity declining; early mid-season; light coloured, soft fruit; high yields; very winter hardy
Redchief	Midwestern US (USDA and AES-Maryland)
	Regionally popular; mid-season; difficult to cap; moderate vigour; good processing quality; resistant to red stele and *Verticillium*
Redcrest	Pacific Northwest (USDA and AES-Oregon)
	Regionally popular; uniform colour; easy to cap; concentrated ripening; good for processing
Reiko	Korea and Japan (Chiba-ken AES-Japan)
	Currently only in Korea; excellent for forcing culture; conical, large and regular fruit; glossy red; very susceptible to powdery mildew
Rhapsody	UK (Scottish Horticultural Research Institute)
	Regionally popular; late; medium to large with red interior, white tips; hard to cap; resistant to red stele and verticillium wilt
Sachinoka	Western Japan (National Research Institute of Vegetables, Ornamental Plants and Tea)

Table 1.3. *Continued*

	Growing in popularity in forcing culture; very firm and good shelf life; high brix and vitamin C
San Miguel	California (Driscoll Associates, California)
	Regionally popular; early; very good size and appearance; medium red; holds size well
Seascape	California/all mild climates (University of California)
	Widely planted; day-neutral; large, firm fruit with bright red internal and external colour
Selva	California/all mild climates (University of California)
	Widely planted; large, firm fruit; somewhat hollow; very productive
Senga Sengana	Poland/Scandinavia (Max-Planck-Institut für Kulturpflanzenzuchtung, Hamburg, Germany)
	Regionally popular; very stress tolerant; low yields; caps easily; excellent for processing
Shuksan	Pacific Northwest (Washington AES)
	Regionally popular; mid-season; good internal colour; good processing quality; some misshapen fruit; resistant to red stele, *Verticillium* and *Botrytis*
Sivetta	The Netherlands (Institute for Horticultural Breeding, the Netherlands)
	Currently little grown; productive in waiting bed culture; low fruit quality
Startyme	Ontario (Horticultural Research Institute of Ontario)
	New release; late; large bright fruit with slightly pale interior; caps easily; vigorous
Suhong	Korea (Pusan Horticultural Experimental Station)
	Widely planted; high vigour; large, bright red fruit; somewhat soft; resistant to anthracnose, powdery mildew and fusarium wilt
Sunrise	Lower midwestern US (USDA-Maryland)
	Currently little planted; early; glossy, firm skin; light red skin and pale flesh; resistant to red stele, verticillium wilt, leaf scorch and mildew
Surecrop	Midwestern US (USDA and AES-Maryland)
	Currently little planted; midseason; very stress adapted; low yields; resistant to multiple leaf diseases, red stele and verticillium wilt
Swede	California (Driscoll Associates, California)
	Currently little planted; large, symmetrical with dark red exterior and light red interior colour; high flavour; only moderately firm; white–green shoulders
Sweet Charlie	Southern US (University of Florida)
	Currently widely planted; early; large, orange–red fruit; high flavour; highly productive; resistant to *Colletotrichum*
Symphony	UK (Scottish Crop Research Institute)
	New release; late mid-season; large and firm with red–crimson

Table 1.3. *Continued*

	colour; flavourful; excellent shelf life; resistant to red stele and crown rot
Tamella	Cool regions of Europe (Institute for Horticultural Breeding, the Netherlands)
	Currently little planted; early; productive in both outdoor and protected cropping; moderate firmness, very susceptible to *Phytophthora cactorum*
Tangi	Southern US (Louisiana-AES)
	Popularity declining; early; red throughout; vigorous; resistant to leaf spot and scorch
Tango	UK (Horticultural Research International, UK)
	Regionally popular; strong day-neutral; very early; modest fruit quality; high productivity; resistant to red stele and verticillium wilt
Tethis	Southern Italy (Consorzio Italiano Vivaisti Co.)
	New release; mid-season with extended season; very large and firm; bright red exterior and medium red interior; produces a second crop
Thomas	California (Driscoll Associates, California)
	Currently little planted; large, symmetrical fruit; short cropping cycle
Tochiotome	Eastern Japan (Tochigi AES)
	Growing in popularity in forcing culture; high yielding; very attractive glossy appearance; good flavour; large and firm fruit
Totem	Pacific Northwest (Department of Agriculture Research Station, British Columbia)
	Currently widely grown; late; caps easy and has good internal colour; darkens; acid flavour; good for processing; resistant to red stele, *Verticillium* and *Botrytis*
Toyonoka	Japan (National Research Institute of Vegetables, Ornamental Plants and Tea)
	Widely grown; currently under PVC tunnels; large and firm fruit; very sweet
Tribute	Midwestern, mid-Atlantic and eastern US (USDA and AES-Maryland)
	Currently widely grown; day-neutral; medium to small fruit; good for processing; resistant to red stele and powdery mildew
Tristar	Midwestern and eastern US (USDA and AES-Maryland)
	Regionally popular; day-neutral; medium to small fruit; good for processing; resistant to red stele and powdery mildew.
Trumpeter	Upper midwestern US
	Currently little grown; late; very winter hardy; medium sized fruit
Tudla	Southern Italy, Spain (Planasa, Spain)
	Regionally popular; early, large elongated berry; irregular; very dark with occasionally dark skin
Tufts	California (University of California)
	Currently little grown; large, firm and attractive fruit; red skin,

Table 1.3. *Continued*

	orange—red flesh; early fruit malformed; June yellows
Veestar	Eastern Canada (Horticultural Research Institute, Vineland, Ontario)
	Regionally popular; very early, medium-sized fruit with light flesh; low yields
Winona	Upper midwestern US (USDA-Maryland and AES-Minnesota)
	New release; late; glossy, scarlet fruit with red—orange centre; winter hardy; resistant to common leaf diseases, red stele and black root rot
Zefyr	Scandinavia (Spangsbjerg, Denmark)
	Regionally popular; early; medium to small fruit, soft, modest quality

new Driscoll releases (T. Sjulin and D. Shaw, California, 1998, personal communication). 'Pajaro', 'Selva' and 'Oso Grande' have been mostly replaced in Florida by 'Sweet Charlie' , 'Key Largo' and 'Camarosa'. The new UC-Davis release 'Gaviota', and 'Captiva' and 'Mirador' from Driscoll Strawberry Associates show high promise. In the southern US, annually grown 'Chandler', 'Sweet Charlie' and 'Camarosa' have made strong inroads into the acreage of a number of older matted row types including 'Earliglow', 'Cardinal', 'Sunrise', 'Atlas', 'Apollo' and 'Tangi'.

In the colder regions of North America, there has been much less change in cultivar composition. In the midwestern and northeastern US, 'Honeoye', 'Kent' and 'Earliglow' have remained important over the last 15 years, with 'Redchief' and 'Blomidan' being replaced by 'Annapolis', 'Jewel', 'Cavandish' and most recently 'Delmarvel'. The new releases, 'Chambly', 'Mira', 'Mohawk', 'Noreaster' and 'Winona' show promise (Galletta *et al.*, 1997; Jamieson, 1997). In eastern Canada and the upper midwestern US, 'Veestar', 'Kent', 'Glooscap' and 'Bounty' have stayed popular, with 'Annapolis' and 'Honeoye' replacing diminishing acreages of 'Redcoat', 'Bounty' and 'Blomidan'. Promising new cultivars in this region are 'Mohawk', 'Mira', 'Mesabi' and 'Startyme'. In the Pacific Northwest, 'Totem' has dominated for the last decade, with 'Redcrest' and 'Hood' replacing 'Benton' and 'Shuksan' as minor cutivars.

The most dramatic shifts in cultivar choice in Europe have also occurred in the warmer climates, although a few areas have remained static. The California cultivar 'Pajaro' still is important in southern Italy, but a number of newer varieties are being widely planted including 'Camarosa', 'Chandler', 'Tudla', and locally bred 'Eris', 'Tethis', 'Clea' and 'Miranda' (Faedi *et al.*, 1997; Tom Sjulin, California, 1998, personal communication). 'Camarosa' shows signs of being the next dominant cultivar (D. Shaw, California, 1998, personal communication). In the Po Valley of northern Italy, 'Marmolada' dominates, followed by 'Idea', 'Addie' and 'Miss' (D. Simpson, UK, 1998, personal communication). 'Elsanta', 'Marmolada', 'Selva' and 'Seascape' are important in the northern mountainous regions (Faedi *et al.*, 1997). In Spain, 'Camarosa' is by far the leading cultivar, with some 'Tudla', 'Oso Grande' and 'Pajaro' being planted (López-Aranda and Bartual, 1999). Many new cultivar candidates are being produced locally (Bartual *et al.*, 1993, 1997; Sanchez Eguialde, 1997), but none have become widely grown. In France, 'Elsanta' and to a lesser extent, 'Gariguette' have remained dominant over the last decade, although 'Elsanta' is now being challenged by 'DarSelect' and 'Eros' in the southwest (D. Simpson, UK, 1998, personal communication). Some 'Pajaro' is still grown in southeastern France, together with minor acreages of 'Chandler' in the Loire Valley. 'Seascape' and 'Selva' remain the dominant day-neutrals in France, but only limited acreage is planted (T. Sjulin, California, 1998, personal communication). There are also a number of new locally bred selections that show promise (Roudeillac and Markocic, 1997).

In the cooler regions of Europe 'Elsanta' and 'Senga Sengana' remain dominant, but new cultivars are beginning to make an impact. In the UK, 'Elsanta' is most important, but 'Symphony' and 'Pegasus' are widely planted and 'Honeoye' has gained a foothold. Three day-neutrals are also popular: 'Evita', 'Bolero' and 'Tango' (McNichol *et al.*, 1997; Simpson *et al.*, 1997a, b; D. Simpson, UK, 1998, personal communication). The new release 'Florence' is attracting the most attention. In Germany, the Netherlands and Belgium, 'Elsanta' also dominates, but some 'Selva', 'Korona' and 'Elvira' are being planted (Meesters and Pitsioudis, 1997). 'Senga Sengana' is still the dominant cultivar in Poland and Scandinavia, but at least in Scandinavia, other cultivars are being widely planted including 'Korona', 'Honeoye', 'Zefyr', 'Bounty' and 'Dania' (Martinsson, 1997; Nes, 1997), and several breeding programmes are producing new possibilities (Hietaranta and Linna, 1997; Trajkovski, 1997). In eastern Europe, 'Gorella' and 'Elsanta' are challenging the traditional leadership of 'Senga Sengana' (Dénes, 1997; Stanisavljević *et al.*, 1997). In Russia and northern Europe, a wide mixture of local bred and imported cultivars are grown. Some of the more popular are: 'Kokinskaja Rannaja', 'Festivalnaja Romashka', 'Zarya', 'Zenit', 'Nadezhda', 'Lurck VIRa', 'Rannyaya Plotnaya', 'Istochnik', 'Desna' and 'Senga Sengana' (Govorova, 1992).

Cultivar usage has changed only gradually in most other major worldwide production regions. In Japan, 'Toyonoko' and 'Nyoho' have remained dominant for over a decade, in spite of over 50 cultivars being released in the last 10 years (Kawagishi, Japan, 1998, personal communication). 'Toyonoko' predominates in west Japan, whereas 'Nyoho' is the leading cultivar in east and south Japan. Among the new cultivars, 'Sachinoka' and 'Tochiotome' are receiving the most interest in these regions. 'Hokowase', which was once grown all over Japan, is now only important in Hokkaido, along with the new release 'Kitaekubo'. In Korea, locally bred 'Suhong' dominates together with 'Reiko' and 'Nyoho'. In Israel, infra-short-day types such as 'Rachel' are replacing traditional acreage of 'Douglas', 'Tufts' and 'Chandler' (Izhar, 1997). In Turkey, California cultivars such as 'Douglas', 'Pajaro' and 'Cruz' now dominate in most areas, replacing old landraces of native *F. vesca* and a mixture of cutivars introduced from various locations across the world (Kaşka, 1997). In northwestern Turkey, 'Senga Sengana' is also grown (D. Simpson, UK, 1998, personal communication). The newer varieties from the University of California breeding programme are being tested at several locations across Turkey and a breeding programme has been initiated at Cukurova University. The California cultivars, 'Chandler', 'Oso Grande', 'Selva', 'Douglas', 'Pajaro' and 'Sequoia' are popular in Egypt, together with 'Sweet Charlie' from Florida, and the Israeli cultivars 'Sharon' and 'Ofira'. 'Camarosa' from California is only beginning to be utilized. In Australia, the leading types are 'Chandler', 'Pajaro', 'Parker' and 'Selva', although a wide array of Australian cultivars are also grown (Zorin *et al.*, 1997). In South America, the primary cultivars are 'Chandler', 'Pajaro', 'Selva', 'Oso Grande', 'Sweet Charlie' and 'Tudla'.

CONCLUSIONS

The greatest concentration of strawberry production is now in the northern hemisphere in Mediterranean climates with mild summer and winter temperatures. Particularly high levels of production are also found in the mild climates of Korea and Japan. The industries in Spain, the Korean Republic and the USA have shown the greatest growth over the last 20 years. It is expected that the world industry will continue to expand, with increased growth in northern Africa, west-central Asia and perhaps South America.

The most widespread cultivars are those developed in California: 'Camarosa', 'Seascape', 'Selva' and 'Chandler'. They dominate all warm climates across the northern and southern hemisheres. Cultivar usage in cold climates is much more regional, with only a few varieties, such as 'Honeoye' and 'Elsanta', achieving international attention. Extremely active breeding programmes are found all across the world, and it is likely that local cultivars will eventually make inroads into the dominance of California types.

2

THE STRAWBERRY SPECIES

THE STRAWBERRY SPECIES

Numerous species of strawberries are found in the temperate zones of the world. Only a few have contributed directly to the ancestry of the cultivated types, but all are an important component of our natural environment. The strawberry belongs to the family *Rosaceae* in the genus *Fragaria*. Its closest relatives are *Duchesnea* Smith and *Potentilla* L. There are four basic fertility groups in *Fragaria* that are associated primarily with their ploidy level or chromosome number. The most common native species, *F. vesca*, has 14 chromosomes and is considered to be a diploid. The most important cultivated strawberry, *F. × ananassa*, is an octoploid with 56 chromosomes. Interploid crosses are often quite difficult, but species with the same ploidy level can often be successfully crossed. In fact, the most commonly cultivated strawberry, *F. × ananassa*, is a hybrid of two New World species, *Fragaria chiloensis* (L.) Duch. and *Fragaria virginiana* Duch. (see Chapter 3).

Although a large number of the strawberry species are perfect flowered, several have separate genders (Table 2.1). Some are dioecious and are composed of pistillate plants that produce no viable pollen and function only as females, and staminate male plants that produce no fruit and serve only as a source of pollen (Fig. 2.1). The perfect-flowered types vary in their outcrossing rates from self-incompatible to compatible (Table 2.1). Isozyme inheritance data have indicated that California *F. vesca* is predominantly a selfing species (Arulsekar and Bringhurst, 1981), although occasional females are found in European populations (Staudt, 1989; Irkaeva *et al.*, 1993; Irkaeva and Ankudinova, 1994). Ahokas (1995) has identified at least two different self-incompatible genotypes of *F. viridis* in Finland.

An accurate taxonomy of the strawberry species is still emerging. Although the European and American species of *Fragaria* have been rigorously defined by Staudt (1962, 1989, 1999), the species situation in Asia is much more ambiguous (Hummer, 1995). There is no overall consensus of

25

Table 2.1. Wild strawberries of the world and their fruiting characteristics (adapted from Galletta and Bringhurst, 1990; Staudt, 1989; Hummer, 1995; Bors and Sullivan, 1998).

Ploidy	Species	Breeding system	Fruit characteristics
2x	F. vesca	Self-compatible[a]	Long ovate-variable, bright red, raised seeds, very aromatic, very soft flesh
	F. viridis	Self-incompatible	Firm, green–pink, aromatic, seeds in pits, spicy, cinnamon-like flavour
	F. nilgerrensis	Self-compatible	Subglobose, pink, tasteless to unpleasant, many sunken seeds
	F. daltoniana	Self-compatible	Ovoid to cylindrical, shiny red, tasteless
	F. nubicola	Self-incompatible	Resembles F. vesca
	F. innumae	Self-compatible	Elongate, achenes sunken, spongy and nearly tasteless
	F. yesoensis	Self-compatible	Resembles F. nipponica
	F. mandshurica	Self-incompatible	Subglobose to obovoid, seeds in shallow pits, very acid
	F. nipponica	Self-incompatible	Globose to ovoid, seeds in pits, unpleasant flavour
	F. gracilisa	Self-incompatible	Elongated and ovate
	F. pentaphylla	Self-incompatible	Ovoid-globose, bright red-coloured, firm but little flavour
4x	F. corymbosa	Dioecious	Seeds in deep pits
	F. orientalis	Trioecious	Soft, obovoid, slight aroma, seeds sunken
	F. moupinensis	Dioecious	Resembles F. nilgerrensis, orange–red, spongy, nearly tasteless
5x, 6x	F × bringhurstii	Dioecious	Intermediate to F. chiloensis and F. vesca
6x	F. moschata	Trioecious	Light to dark dull purplish red, soft, irregular to ovoid, musky or vinous, aromatic, raised seeds
8x	F. chiloensis	Trioecious	Dull red–brown, white flesh, mild, firm, round to oblate, small to large, achenes raised or sunken
	F. virginiana	Trioecious	Twice the size of F. vesca, soft, light to deep red or scarlet, white flesh, tart, aromatic, raised achenes, strongly reflexed calyx
	F. iturupensis	Trioecious	Spherical, bright red, superficial achenes
	F. × ananassa	Trioecious	Very large, red, variable in all traits

[a] Occasional females are found in European populations.

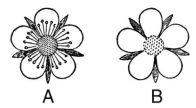

Fig. 2.1. A perfect or hermaphrodite strawberry flower (A) having both pistils and stamens, and a pistillate flower (B) having pistils but no stamens (from Fletcher, 1917).

what species exist in China and published descriptions have sometimes relied on limited collections. The discussions here rely on Staudt's (1989) last worldwide treatment of *Fragaria*, with the addition of two Chinese diploids recently described by Yu *et al.* (1985).

The species recognized by Staudt (1989, 1999) and Yu *et al.* (1985) can be organized according to their geographic range (Fig. 2.2). *Fragaria vesca* has the largest native range, encompassing most of Europe, Asia, and the Americas. The rest of the species are more restricted in ecogeography, being clustered primarily in Euro-Siberia, northern China and Manchuria, Indo-South China, Japan, and the Americas. Japan is particularly species rich, with at least four endemic species radiating across its islands. The cultivated strawberry, *F. × ananassa*, is grown in almost all arable zones of the world, although its native range is restricted to the Pacific Northwest of North America.

STRAWBERRY SPECIES DISTRIBUTIONS

Worldwide

Fragaria vesca L. (2*n* = 2*x* = 14)
The wood or alpine strawberry, is cultivated to a limited extent in North America and Europe. It has thin light-green, sharply serrated leaves borne on slender petioles (Fig. 2.3). The terminal tooth of the terminal leaflet is usually longer than the adjacent lateral teeth and the calyx is reflexed. The plant is erect and 15–30 cm tall. Flowers are bisexual, approximately 1.3 cm wide; inflorescences are about the same length or taller than the leaf petioles. Most plants are short day, but everbearing types exist (*F. vesca* f. *semperflorens*). Fruits are long ovate, bright red in colour and highly aromatic. The fruit has very soft flesh and raised or superficial seeds. Runnerless and white fruited forms exist.

There are four subspecies found in the group (Staudt, 1962, 1999): (i) *F. vesca* ssp. *vesca* – woods of Europe and Asia; (ii) ssp. *americana* (Porter) Staudt – woods of eastern North America to British Columbia), 3) ssp. *bracteata* (Heller) Staudt – woods of western North America; and (iv) ssp. *californica* (Chamisse

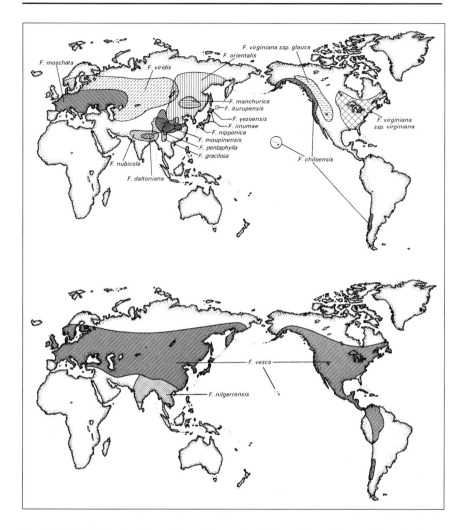

Fig. 2.2. World distribution of *Fragaria* species (adapted from Hancock and Luby, 1993).

and Schlechtendal) Staudt – California. Several ecotypes have been described within ssp. *californica* including headland scrub, coastal forest and Sierran forest (Table 2.2). All of these subspecies are hermaphroditic and self-fertile, except for ssp. *braceata* which has both hermaphrodites and occasional females (Staudt, 1989).

Fragaria × *ananassa* L. ($2n = 8x = 56$)
This is now the most important strawberry cultivated worldwide, but its domestication was not based on natural hybrids between *F. chiloensis* and

Fig. 2.3. Duchesne's drawing of *Fragaria vesca* (from Darrow, 1966). Cytogenetic studies suggest that this species may be a diploid progenitor of the octoploid strawberries.

F. virginiana, but instead on accidental hybrids that appeared in European gardens in the mid-1700s (see Chapter 3). From a horticultural point of view, many of the traits distinguishing the two species are complementary (Table 2.3), and it is not surprising that hybrid derived populations came to dominate commercial plantings of strawberry. Even after dozens of rounds of selection, many of the morphological traits found in *F. × ananassa* are still intermediate to its parent species, but considerable segregation has occurred. Using morphological traits, Darrow (1966) found that eastern cultivars expressed 27–57% of the characters from *F. chiloensis*.

Euro-Siberia

Fragaria viridis Duch. (2n = 2x = 14)

This is a slender, upright plant with dark green leaves with smaller serrations than *F. vesca* (Fig. 2.4). It is found in open grassland hills, steppes, at the edge of forests and among brush. It produces only a few nodeless runners. Flower numbers per inflorescence are smaller than *F. vesca*, but it has perfect flowers that are larger than *F. vesca*. The petals overlap and are often yellowish-green when opening. Fruit is small but larger than *F. vesca*, firm, green to pink in

Table 2.2. Ecotypes of *Fragaria vesca* and *Fragaria chiloensis* found in California (Hancock, 1990).

| Species | Environment | Mean rainfall (mm year^{-1}) | Mean temperature (°C) | | Soil characteristics | | Salinity | |
			January	July	Type	% carbon	ppm	pH
Fragaria vesca	Headland scrub	686	10.1	15.9	Silt loam	2.8	497	6.0
	Coastal forest	1160	8.9	16.7	Silt loam	5.9	326	6.3
	Sierran forest	1453	6.5	17.2	Silt loam	5.5	205	5.6
Fragaria chiloensis	Dunes	476	10.4	10.5	Sand	0.3	663	6.7
	Coastal strand	1837	7.7	8.1	Sand	0.3	652	6.7
	Headland scrub	482	9.5	10.3	Silt loam	2.8	497	6.0
	Woodland meadows	1416	7.9	8.2	Sandy loam	2.8	390	5.4

Table 2.3. The characteristics that separate *F. chiloensis* from *F. virginiana* (Darrow, 1966).

Character	*F. chiloensis*	*F. virginiana*
Leaves	Thick, leathery	Thin
	Deep set stomata	Shallow set stomata
	Strongly netted	Not strongly netted
	Glossy	Dull
	Dark green	Bluish to light green
	Evergreen	Deciduous
	Short teeth	Coarse teeth
Petioles	Thick	Slender
	Not channelled	Broadly channelled
Runners	Robust	Slender
	Red	Green
	Persistent	Ephemeral
Crown	Thick	Less thick
Flowers	Large	Small
	Large stamens	Small stamens
Fruit	Dull red	Scarlet to crimson
	Seeds slightly sunken	Seeds sunken in pits
	Late ripening	Early ripening
	Large	Small
	Firm	Soft

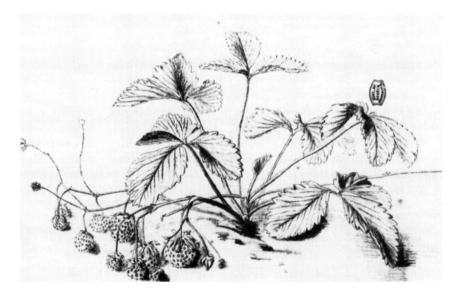

Fig. 2.4. Duchesne's drawing of *Fragaria viridis* (from Darrow, 1966).

colour, and aromatic. The scapes lie along the ground when the berries are ripe. Seeds are set in pits. The calyx is clasping and hard to separate. *F. viridis* can be distinguished from *F. vesca* by its phosphoglucose isomerase isozyme pattern (Arulsekar and Bringhurst, 1981).

Fragaria orientalis Losinsk ($2n = 4x = 28$)
This is a small, upright plant with long, slender runners. It is found in forests and open mountain slopes. Its leaves are ovate, light green, nearly sessile with deeply serrate margins. There are a few large flowers (2.5–3 cm) on the inflorescence. Fruit is large, obovoid, and only slightly aromatic. Seeds are sunken.

Fragaria moschata Duch. ($2n = 6x = 42$)
The musky strawberry, is a dioecious, tall, vigorous plant that produces few runners. It grows in forests, under shrubs and in tall grass. Leaves are large, dark green, rugose, rhombic, prominently veined and pubescent. The flowers are large (20–25 cm in diameter) and the inflorescence emerges above the foliage, but due to the weight of the ripe berries, the scapes lie along the ground. The calyx is usually reflexed. The fruit is light red to dull brownish to purplish red, soft, irregular-globose to ovoid, and has a strong vinous flavour. The fruit is slightly larger than that of *F. vesca* and bears raised achenes. The calyx is strongly reflexed. Both white and red, perfect-flowered forms are cultivated to a limited extent under the names Hautboy or Hautbois.

Indo-South China

Fragaria daltoniana J. Gay ($2n = 2x = 14$)
This species is located in a narrow region of the Sikkim Himalayas at elevations of 3000–4500 m. Plants are vigorous, with petiolulate leaves that have few teeth along the margins. Runners are slender and flowers are solitary and self-compatible (Bors and Sullivan, 1998). Fruit range from ovate to cylindrical, are relatively long (2–2.5 cm), bright red, spongy and tasteless. It has shiny, coriaceous leaves.

Fragaria nilgerrensis Schlecht. ($2n = 2x = 14$)
This is vigorous and spreading with pubescent, dark-green and heavily veined leaves. The petioles and peduncles are covered with long, stout hairs. The leaflets are petiolate, round to ovate, with small serrations, dull green and very pubescent. It produces a small inflorescence with three or four large bisexual flowers. The flowers have a pink blush. The fruit are small, subglobose, pale-pink, tasteless to unpleasant and have many small, sunken seeds. Staudt (1989) describes it as having a banana-like flavour. There is a subspecies *hayatai* from Taiwan that has anthocyanin in all parts of the plant, including the berries.

Fragaria nubicola Lindl. ex Lacaita ($2n = 2x = 14$)
This is also native to the temperate Himalayas at 1500 to 4000 m elevation. The plants and fruit closely resemble *F. viridis*, except that they have sympodial runners.

North China — Manchuria

Fragaria gracilisa Lozinsk ($2x = 2n = 14$)
This is found on grassy mountain slopes, ditches and in forests of Shanxi, Guanchu, Qihai, Henen, Hubei, Sichuan, Yunnan and Tibet. It is tri- to penni-form five foliate with adpressed soft hairs, and lanceolate sepals that reflex upon fruit ripening. They are self-incompatible (Bors and Sullivan, 1998) and the fruit are elongated, ovoid.

Fragaria pentaphylla Lozinsk ($2x = 2n = 14$)
This is found in grassy mountain slopes at 1000 to 2000 m elevation in Shanxi, Guanshu and Sichun. It has penniform, thick, five-foliate leaves, reflexed sepals and elongated calycules at fruit maturity. They are self-incompatible (Bors and Sullivan, 1998) and the fruit is ovoid–globose.

Fragaria mandschurica Staudt ($2x = 2n = 14$)
This is native to Manchuria. It closely resembles the autotetraploid *F. orientalis*, except that the flowers are smaller and the leaves and teeth are less coarse than *F. orientalis* (Staudt, 1989). The berries are subglobose to obovoid with mostly yellow achenes in shallow pits. The berries are very acid.

Fragaria corymbosa ($2x = 4n = 28$)
This species was described by Staudt (1989) from a limited collection of a male plant from north China collected by Losina-Losinskaja (Staudt, 1989). It is not clear, if this specimen represents a native population (Hummer, 1995) or how it compares with the two diploid Chinese species described by Yu *et al.* (1985). The only published information about it is that the leaflets are somewhat oval, and from crossing experiments, Staudt deduced that its achenes sit in deep pits.

Fragaria moupinensis (French.) Card. ($2x = 4n = 28$)
The plants and fruit of this species are very similar to *F. nilgerrensis*. The leaves are trifoliate, serrate elongated oval, with the lower leaflets being smaller. The inflorescence is longer than the leaf petioles and has only two to four flowers. Runners are short. The fruit are orange–red coloured, with deeply set achenes and the flesh is spongy and nearly tasteless.

Japan

Fragaria iinumae Makino ($2x = 2n = 14$)

This species is restricted to the alpine mountains of central and northern Japan. It is a vigorous, erect plant with slender filiform runners. Leaflets are subglaucous in colour, broadly obovate or cuneate-orbicular, rounded at the apex, petiolate with margins that are coarsely dentoserrate. They are glabrous above with appressed to ascending long pubescence beneath especially on the nerves. Only a few scapes are produced that are one- to three-flowered. Flowers are 15–25 mm across, have more than five petals and are self-incompatible (Bors and Sullivan, 1998). The fruit is elongate, 8 mm across × 1.5 cm long with a small calyx, and sunken achenes. The fruit are spongy and nearly tasteless. *F. iinumae* appears to be deciduous, as no leaves are visible during the winter. The glaucous leaf of *F. iinumae* is unique to the rest of the diploids.

Fragaria yesoensis Hara. ($2x = 2n = 14$)

This species is found together with *F. iinumae* in cooler regions of northern Japan and closely resembles *F. nipponica* except that it has spreading hairs (Ohwi, 1965). Terminal leaflets are rhombic–obovate, rounded at the apex, dentate–serrate with silky pubescence on both sides. The scapes are up to 15 cm long and the flowers are 1.5–2 cm across. The plants are deciduous, going completely dormant in winter.

Fragaria nipponica Lindl. ($2n = 2x = 14$)

This is found in the mountains of Japan. It is thought to be closely allied to *F. yesoensis* (Ohwi, 1965). Terminal leaflets are elliptic to broadly ovate with ovate or subdeltoid teeth, pale green colour and appressed pubescence especially on nerves beneath. Scapes are 2–2.5 cm across and have one to four flowers. The fruit is globose to ovoid (1.5–3 cm across), with an unpleasant taste, and its achenes are within pits. Staudt (1989) suggests that there is an undescribed species in the Himalayas that is very similar to *F. nipponica*.

Fragaria iturupensis Staudt ($2n = 8x = 56$)

This is the only octoploid species located in Asia and it is found solely on Iturup Island, northeast of Japan (Staudt, 1989). It has obovate sub-glaucous leaves that are blueish, much like *F. iinumae*. The petiole is covered with patulate hairs. The flowers are hermaphroditic, 16–20 mm wide with 5 petals. There are two to four flowers to an inflorescence. The fruit is similar to *F. vesca*, but larger. Stolons are branched with no secondary runners from axils of primary bracts. Berries are subspherical, bright red and shiny with reflexed calyxes and superficial achenes.

The Americas

Fragaria chiloensis (L.) Duch., ($2n = 8x = 56$)

The beach or Chilean strawberry was once extensively cultivated in western South America and France, but is now only grown to a limited extent (see Chapter 3). Plants are low-spreading and vigorous with prolific runnering (Fig. 2.5), and tend to be evergreen. Flowers are large, 20–35 mm in diameter. Leaves are generally thick, strongly reticulate-veiny beneath, dark-green and very glossy. Runners are robust and bright red. Native forms have fruit which is dull to bright red in colour, with white flesh and mild to pungent flavour. Achenes are reddish-brown to dark brown. Many of the cultivated forms are albino. Fruit is round to oblate with raised or sunken achenes. Fruit size in the cultigens can be in excess of 10 g, but most native forms average 1–3 g.

Wild populations of *F. chiloensis* are either dioecious, gynodioecious or perfect flowered depending on geographical location. North American *F. chiloensis* are primarily dioecious, with staminate plants being about 10% more common than pistillate (Hancock and Bringhurst, 1979b, 1980). In some cases, apparent males are polygamodioecious and bear a few early fruit. Highly fertile hermaphrodites have been found in California at Año Nuevo and Pigeon Point, Alaska, and in the northern islands off the coast of British Columbia. In Chile, *F. chiloensis* is largely gynodiecious as all wild plants are either pistillate or hermaphroditic (Lavín, 1997). Plants in Hawaii are all hermaphroditic.

Fig. 2.5. Duchesne's drawing of *Fragaria chiloensis* (from Darrow, 1966). This species is one of the progenitors of the cultivated species, *Fragaria × ananassa*.

There are four subspecies of *F. chiloensis* recognized (Staudt, 1989): (i) ssp. *lucida* (E. Vilmorin ex Gay) Staudt – coast of Pacific ocean from Queen Charlotte Island to San Luis Obispo, California, (ii) ssp. *pacifica* Staudt – coast of Pacific ocean from Aleutian Islands to San Francisco, California, (iii) ssp. *sandwicensis* (Degener and Degener) – Hawaii, and (iv) ssp. *chiloensis* (L.) Duch. – beaches and mountains of South America. Two forms of this subspecies are recognized, the cultivated f. *chiloensis* and the native f. *patagonia*. Recent morphometric and random amplified polymorphic DNA (RAPD) analyses of interspecific variation in *F. chiloensis* have indicated that ssp. *lucida* and *pacifica* might intergrade too much to be considered separate subspecies, but ssp. *sandwicensis* and *chiloensis* are distinct (Porebski and Catling, 1997; Catling and Porebski, 1998). The major characteristics used to separate the subspecies were hair length, leaflet size, plant colour, petal number and whether the hairs on the leaf stalk were ascending or spreading. Hair orientation was the only reliable way to distinguish ssp. *lucida* from *pacifica*.

Several ecotypes of *F. chiloensis* have been identified in both North and South America. Distinct dune, coastal strand, headland scrub and woodland meadow types are found in California (Table 2.2). They are distinguished primarily by flower number, leaf width, leaf biomass, runner width and resistance to salt and drought stress. The woodland-meadow types may be stabilized hybrid derivatives of *F. chiloensis* × *F. virginiana* (Hancock and Bringhurst, 1979b). At least two distinct native races have been described in Chile, a coastal type with dark, more glossy, green leaves, and a higher elevation form with duller leaves and a blue casting, much like *F. virginiana* ssp. *glauca* (Cameron *et al.*, 1993). In a morphometric analysis of Chilean *F. chiloensis*, del Pozo *et al.* (unpublished) identified four cluster groups among wild accessions, although they did not specify any climatic or regional patterns to the variability. The most diagnostic characteristics were leaflet size, plant size, weight of fruit and fruit size. Interestingly, white forms of native *F. chiloensis* were discovered that clustered very closely to the white, much larger-fruited domesticated forms (see Chapter 3).

Native hybridizations between *F. vesca* and *F. chiloensis* in coastal California have resulted in persistent $5x$, $6x$ and $9x$ colonies (Fig. 2.6) (Bringhurst and Senanayake, 1966). These have been named *Fragaria* × *bringhurstii* after their discoverer R.S. Bringhurst (Staudt, 1989). Their leaves are intermediate between *F. chiloensis* and *F. vesca* with regard to thickness, colour, profile, pubescence and the appearance of the upper leaf surface. They are mostly sterile at $2n = 35$, 42 or 63, but small percentages of aneuploid gametes are produced that are interfertile with octoploid material.

F. virginiana Duch. ($2n = 8x = 56$)

The scarlet or Virginia strawberry, is found in meadows throughout central and eastern North America. Plants are slender, tall and profusely runnering (Fig. 2.7). Leaves are coarsely toothed and obovate to oblong. The terminal

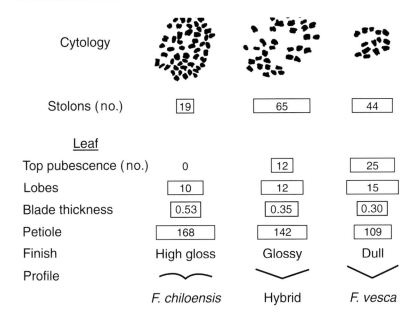

	F. chiloensis	Hybrid	F. vesca
Cytology			
Stolons (no.)	19	65	44
Leaf			
Top pubescence (no.)	0	12	25
Lobes	10	12	15
Blade thickness	0.53	0.35	0.30
Petiole	168	142	109
Finish	High gloss	Glossy	Dull
Profile			

Fig. 2.6. The morphological and cytogenetic traits distinguishing *F. chiloensis*, *F. vesca* and their pentaploid hybrid at Point Sur, California (adapted from Bringhurst and Khan, 1963).

Fig. 2.7. Duchesne's drawing of *Fragaria virginiana* (from Darrow, 1966). This species is the North American progenitor of the cultivated strawberry, *F. × ananassa*.

tooth of the terminal leaflet is usually shorter than the adjacent lateral teeth. The inflorescence is variable, basal to high branching and flowers are small to large (0.6–2.5 cm) and dioecious. The fruit is soft, and round, up to 3 cm in diameter (although most fruit is much smaller). It is light red with white flesh, aromatic and has deeply embedded seeds. *F. virginiana* can be distinguished from *F. chiloensis* by a number of morphological traits (Table 2.3).

Populations of *F. virginiana* vary from being completely dioecious to trioecious. In western populations of *F. virginiana*, all three sexes are found in similar proportions, whereas in eastern populations, only hermaphrodites and females are observed, again in relatively equal proportions (Staudt, 1968; Luby and Stahler, 1993). Levels of fertility in hermaphrodites are highly variable, with large ranges of fertility being found in natural populations from pure males to pure hermaphrodites (Hancock and Bringhurst, 1979b, 1980; Stahler, 1990). In general, females are more fertile than hermaphrodites (Dale *et al.*, 1992). Environment probably plays a role in the fertility of hermaphrodites (Stahler, 1990; Dale *et al.*, 1992), but most of the control is genetic. The level of fertility in hermaphrodites appears to be polygenic as large ranges of fertility are generated in crosses of native males with cultivated hermaphrodites (Scott *et al.*, 1962a) and fruit set in native hermaphrodites can be greatly enhanced through recurrent selection (Luby *et al.*, 1995).

F. virginiana has undergone considerable genetic differentiation like the other octoploid species, *F. chiloensis*. Staudt (1962) described four subspecies: (i) ssp. *glauca* (Wats.) Staudt – southern Rocky mountains to northwest Canada and central Alaska; (ii) ssp. *platypetala* (Rydb.) Staudt – Rocky mountains from Wyoming to Colorado, west to Sierra Nevada and Cascades Mountains; (iii) ssp. *grayana* (E. Vilmorin ex Gay) Staudt – meadows and forest margins from New York to Alabama, Louisiana and Texas; and (iv) ssp. *virginiana* Duch. – meadows and forest margins from eastern United States, Newfoundland and Yukon Territory. Staudt felt that ssp. *virginiana* was more common in the north and ssp. *grayana* more common in the south.

The western subspecies have yielded considerable taxonomic debate. Darrow (1966) considered the western types of *F. virginiana* to be a separate species, *F. ovalis*, but Staudt (1962) did not give species rank to this material because of the lack of barriers to hybridization and the intermediacy of its characters. The separation of ssp. *platypetala* and ssp. *glauca* has also been questioned. Welsh *et al.* (1987) suggested that the two subspecies completely intergrade and probably should be referred to as a single taxa var. *glauca*. Others have followed this designation (Scoggan, 1978) or have not attempted to recognize infraspecific taxa (Dorn, 1984). What was formerly recognized as a distinct species, *F. multicipita*, was recently reduced to *F. virginiana* ssp. *glauca* forma *multicipita* when it was discovered that its unique morphology was caused by a phytoplasma (Catling, 1995).

Whereas *F. virginiana* ssp. *virginiana* and ssp. *glauca* appear to be largely distinct across their range, strawberries in the Black Hills and eastern front ranges of the Rocky Mountains may be introgressive swarms between ssp. *glauca* and ssp. *virginiana* (Hokanson *et al.*, 1993; Luby *et al.*, 1992; Sakin *et al.*, 1997). In a mutivariate analysis of *F. virginiana and F. chiloensis* populations across the northern USA, Harrison *et al.* (1997b) found that Black Hill populations are intermediate between collections of eastern ssp. *virginiana* and western ssp. *glauca* for morphological traits, and when analysed with RAPD markers, the Black Hill populations were part of a large cluster group incorporating both ssp. *virginiana* and *glauca* (Fig. 2.8). The Black Hill populations

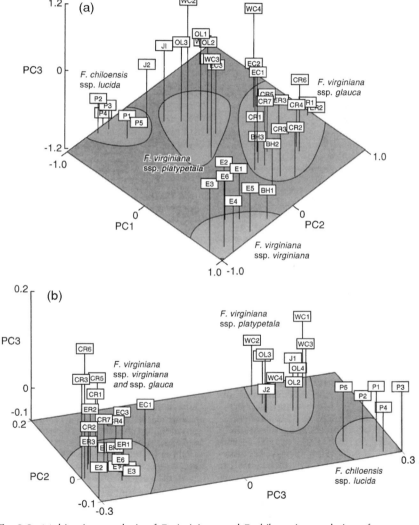

Fig. 2.8. Multivariate analysis of *F. virginiana* and *F. chiloensis* populations from across the USA (Harrison *et al.*, 1997b). The factor scores are for the first three principal components of the means of morphological variables (a) and RAPD frequencies (b). Groups are labelled according to the *Fragaria* taxonomy of Staudt (1962).

probably represent relics of the late Pleiocene when the Great Plains were mainly a boreal forest that provided a continuous habitat for hybridization between eastern and western forms of *F. virginiana*.

Natural hybrids of *F. virginiana* × *F. chiloensis* exist where native habitats of the two species overlap in British Columbia, Washington, Oregon and northern California. Staudt (1999) has designated these hybrids as *F.* × *ananassa* ssp. *cunefolia* (Nutt ex Howell). The *F. chiloensis* populations noted on woodland–meadow sites by Jensen and Hancock (1981) probably represent stabilized hybrid derivatives of this cross. Hybrid populations of *F. chiloensis* × *F. virginiana* are common from Vancouver Island along the coast to Fort Bragg, California and Staudt (1989) has indicated that 'the further one goes from the coastal area the more the *F. chiloensis* characters decrease. Plants with somewhat thinner leaves but some other characters of *F. chiloensis* are combined in ssp. *platypetala* of *F. virginiana* . . . considered to be the final link of introgression of *F. chiloensis* into *F. virginiana* ssp. *glauca*.' Luby *et al.* (1992) present evidence of interaction between these two species in the mountains of northern Idaho and western Montana, where individuals of *F. virginiana* have thick, roundish leaves and thick runners reminiscent of *F. chiloensis* even though they are more than 400 km from the Pacific Ocean.

In the recent study by Harrison *et al.* (1997b), variation patterns in the morphological traits suggested that *F. virginiana* ssp. *platypetala* is distinct from *F. chiloensis* ssp. *lucida* in the Pacific Northwest (Fig. 2.8). However, in the RAPD analysis, the two groups were combined even though they remained distinct from all the eastern *F. virginiana*. It is possible that the RAPD markers are selectively neutral traits that reflect ancient patterns of gene flow between *F. chiloensis* and *F. virginiana* in early post-Pleistocene times, whereas the morphological traits were moulded through selection as the genus *Fragaria* faced new environmental challenges.

EVOLUTIONARY RELATIONSHIPS

It is difficult to accurately reconstruct the phylogeny of *Fragaria*, as most species collections are extremely limited. This has greatly hampered cross-ability studies, cytogenetic analyses and molecular comparisons. In many cases, what information we have is based on a single or a few representatives of a species. Hypotheses are made below, but rigorous conclusions on evolutionary relationships await the acquisition and study of larger collections of each species, particularly those of Asian origin.

Although there appear to be some barriers to interfertility among the diploid strawberries, they can all be crossed to some extent, and meiosis in the hybrids is regular, even in cases where the interspecific hybrids are sterile (Federova, 1946; Staudt, 1959; Fadeeva, 1966). This suggests that they may share the same genome, with only cryptic structural differences. Iwatsubo and Naruhashi (1989) found that the chromosomes of *F. nipponica* and *F. vesca* are very similar in morphology, although *F. iinumae* had some distinguishing features. It seems likely that *F. vesca* is ancestral to all the diploids, as its geographical range overlaps or touches almost all the other diploid species and it has been successfully crossed with most of them, including *F. nilgerrensis*, which is sexually isolated from all the other species tested (Fadeeva, 1966).

There appear to be at least three overlapping groups of species that are interfertile (Bors and Sullivan, 1998): (i) *F. vesca*, *F. viridis*, *F. nubicola* and *F. pentaphylla*; (ii) *F. vesca*, *F. nilgerrensis*, *F. daltoniana* and *F. pentaphyta*; (iii) *F. pentaphyllta*, *F. gracilis* and *F. nipponica*. *F. iinumae* may belong in group 3 or in an additional group, as no fertile seeds have been recovered when it was crossed with either *F. vesca*, *F. viridis* or *F. nubicola*, but it has not been crossed with enough other species to accurately classify it. *F. iinumae* does, however, have a glaucous leaf trait that is unique among the diploids, and its chloroplast restriction fragment length polymorphisms (RFLPs) cluster it with *F. nilgerrensis* in a group that is isolated from the rest (Harrison *et al.*, 1997a).

Polyploidy in Fragaria probably arose through the unification of $2n$ gametes, as several investigators have noted that unreduced gametes are relatively common in *Fragaria* (Scott, 1951; Islam, 1960; Bringhurst and Gill, 1970). Staudt (1984) observed restitution in microsporogenesis of a F_1 hybrid of *F. virginiana* × *F. chiloensis*. In a study of native populations of *F. chiloensis* and *F. vesca*, Bringhurst and Senanayake (1966) found frequencies of giant pollen grains to be approximately 1% of the total. Over 10% of the natural hybrids generated between these two species were the result of unreduced gametes.

F. orientalis is likely of autopolyploid origin, with a species similar to *F. viridis* as its progenitor. These two species cross relatively easily and the resulting hybrids are partially fertile (Federova, 1946). Hybrids of *F. orientalis* and $2x$ *F. vesca* are much more difficult to make, although Staudt (1952) was able to produce a fertile hexaploid hybrid between them, and Bors and Sullivan (1998) found the cross of $4x$ *F. vesca* and *F. orientalis* to be relatively easy to make. The origin of the other tetraploids has not been investigated cytogenetically, although all the tetraploids are interfertile (Evans, 1974, 1977; Sebastiampillia and Jones, 1976; Bors and Sullivan, 1998). The diploid *F. mandscurica* is morphologically similar to *F. orientalis* (Staudt, 1959), and *F. moupinensis* looks similar to the diploid *F. nilgerrensis* (Darrow, 1966).

Although there may not be sufficient differentiation among the diploids to warrant the designation of separate genomes at the diploid level (Staudt, 1959), cytogenetic studies have indicated that there are distinct sets of chromosomes associating in the hexaploid and octoploid species (Federova, 1946; Senanayake and Bringhurst, 1967). The inheritance of a few isozyme loci (Arulsekar *et al.*, 1981) and another regulating gender (Ahmadi and Bringhurst, 1991). indicate that chromosome behaviour in the octoploids may be diploidized, although this conclusion awaits more comprehensive inheritance data. If the octoploids are indeed diploidized, strict disomic pairing may have evolved subsequent to polyploidization, as the genomic divergence between the diploid species is probably insufficient to prevent at least some heterogenetic pairing.

F. moschata could have originated from a number of possible interspecies hybridizations, as it crosses readily with *F. viridis* (Schiemann, 1937), *F. nubicola* (Ellis, 1958), *F. nipponica* (Lilienfeld, 1933), *F. orientalis* (Federova, 1946), and with difficulty with *F. vesca* (Mangelsdorf and East, 1927). Dogadkina (1941) suggested that all three of the genomes of *F. moschata* are mostly homologous, but originating from different related species. Ichijima (1930) described two types of chromosome sets in *F. moschata,* two being dumbbell shaped, and another spherical. Schiemann (1937) described plants that looked like *F. orientalis,* that were derived from a pentaploid F_2 population of *F. vesca* × *F. moschata.* Based on morphology, Staudt (1959) suggested that *F. moschata* could have been produced by a cross of *F. vesca* × *F. nubicola,* or *F. orientalis* × *F. vesca.* Federova (1946) hypothesized that the genome of *F. moschata* is *AAAABB,* with the *A* genome being contributed by *F. orientalis,* and the *B* from some species other than *F. vesca.* Lilienfeld (1933) suggested that the other genome was contributed by *F. nipponica.*

Cytogenetic studies indicate that at least two pairs of genomes are represented in the octoploid species. When they are crossed with the diploids *F. vesca* or *F. viridis,* bivalent or multivalent numbers approaching 14 are commonly observed in pentaploid hybrids, suggesting that there is pairing between one set of diploid and octoploid chromosomes, and another set of octoploid chromosomes (Ichijima, 1926, 1930; Federova, 1946; Senanayake and Bringhurst, 1967). An additional set of chromosomes is left as largely univalents, either due to non-homology with the other sets or competition with a homologous set of chromosomes from the diploid. Similar results have commonly been obtained whether *F. chiloensis, F. virginiana* or *F.* × *ananassa* was used as the octoploid parent, although a few studies have reported much higher numbers of bivalents (21–28) in diploid × *F.* × *ananassa* crosses (Yarnell, 1931; Ellis, 1962).

Three genome formulas have been suggested for the octoploids, *AAAABBCC* (Federova, 1946), *AAA'A'BBBB* (Senanayake and Bringhurst, 1967) and *AAA'A'BBB'B'* (Bringhurst, 1990). Federova suggested that the *A* genome came from an ancestor of *F. orientalis* other than *F. vesca,* the *B* from *F. nipponica* and the *C* from *F. vesca.* After examining the pairing relationships of octoploids crossed with $2x$ and $4x$ *F. vesca ,* Senanayake and Bringhurst suggested that the genomic formula should be *AAA'A'BBBB,* as higher bivalent numbers were observed in hexaploid than pentaploid hybrids. This indicated that the chromosome of *F. vesca* had at least partial homology with another set of octoploid chromosomes. Their guess was that the *A* genome was contributed by either *F. vesca* or *F. viridis,* and they had no idea about the origin of the *A'* and *B* genomes. Bringhurst (1990) later suggested that the genome formula should be *AAA'AA'BBB'B'* to reflect his contention that the octoploids are completely diploidized with strict disomic inheritance (Fig. 2.9).

It seems likely that species similar to *F. vesca* and *F. viridis* are in the background of the octoploid strawberries, as chromosomes from both pair regularly with those of *F. chiloensis*, *F. virginiana* and *F. × ananassa*. These two species could actually represent the *A* and *A'* genomes of Senanayake and Bringhurst, since they have undergone some chromosomal divergence, but not enough to prevent the formation of a significant number of bivalents. Using RAPD markers, Davis and Yu (1997) demonstrated cryptic chromosomal variation between these two species. *F. nubicola* or *F. pentaphylla* could also be represented in the *A'* group as they are interfertile with *F. viridis* and can be crossed with at least limited success to *F. vesca*. Indeed, *F. nubicola* has been shown to have residual homology with octoploid genomes, although its chromosomes do not pair as well as those of *F. vesca* (Byrne and Jelenkovic, 1976). It is also possible, that the *A* genome donor of the octoploids is *F. orientalis*, as the chromosomes of these species pair well (Federova, 1946), although *F. orientalis* itself may be derived from species similar to *F. viridis* or

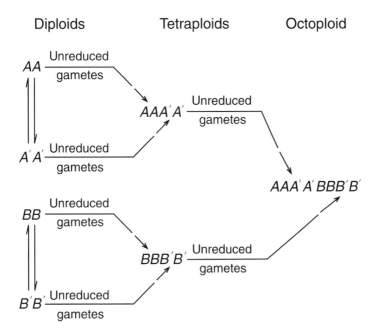

Fig. 2.9. A proposed genomic origin for the octoploid strawberry species (Bringhurst, 1990). Different letters represent genomes from highly divergent species, while those distinguished with a prime originated from much closer relatives.

F. vesca. Perhaps an ancient hybrid of *F. viridis* × *F. vesca* played a role in the formation of the tetraploids or octoploids, as natural hybrids between these species do occur in nature. Accurate resolution of where the octoploid genomes originated awaits future molecular comparisons.

The origin of the other chromosome sets is still very clouded, although we can make some educated guesses. It is possible that the source is *F. nipponica*, but its interfertility with *F. viridis* makes it more likely to represent the *A'* genome group. *F. innumae* is a likely candidate, as it has a unique glaucous leaf colour that is shared by the octoploids *F. iturupensis* from Japan and *F. virginiana* ssp. *glauca* from North America. Hybrids of *F. iinumae* have been produced with *F.* × *ananassa* that were highly fertile after chromosome doubling, suggesting the two species share high levels of chromosome homology (Noguchi *et al.*, 1997). It is also possible that *F. nilgerrensis* is a genomic donor as it can be incorporated into the octoploid genome through artificial crosses (Yarnell. 1931; Bors and Sullivan, 1998; Noguchi *et al.*, 1997). Since *F. nilgerrensis* appears to be the most distantly separated species of the diploids, it is also a good candidate to have contributed a distinct genome. *F. gracilis* and *F. pentaphta* can also not be excluded as possible octoploid genome donors, although cytogenetic studies have not been performed between them and the octoploids. Like *F. vesca*, *F. pentaphyta* is interfertile with a wide number of species, making it possible that it had a pivotal role in the origin of the octoploids (Bors and Sullivan, 1998).

The origin of the North American octoploids also remains clouded. One possibility is that *F. chiloensis* and *F. virginiana* are extreme forms of the same biological species, separated before the Pleistocene (Staudt, 1999), which subsequently evolved differential adaptations to coastal and mountain habitats. The two species are completely interfertile and Harrison *et al.* (1997a) found that they carry similar cpDNA restriction fragment mutations. Nucleotide sequences of the nuclear internal transcribed spacer (ITS) region and two contiguous non-coding regions of cpDNA (the *trnL* intron and the *trnL-trnF* spacer) support a sister relationship between octoploid *F. chiloensis* and *F. virginiana* that may represent a monophyletic origin (Potter *et al.*, 1997). The octoploid strawberries may have originated in east Asia and dispersed along the northern Pacific Islands, through Alaska, and south throughout North America. This would be consistent with the bulk of the diploid species being Eurasian. In fact, the only octoploid located in the eastern hemisphere, *F. iturupensis*, is found in the western Pacific (Staudt, 1973).

It is also possible that *F. chiloensis* and *F. virginiana* have separate origins, with *F. iinumae* playing a more important role in the development of *F. virginiana*, and *F. nilgerrensis* or some other species being more involved in the background of *F. chiloensis*. The cytoplasms of the two octoploid species may be similar because they share a common mother species. In this scenario, *F. virginiana* could have developed from *F. iturupensis* since they have a similar unique leaf colour, whereas *F. chiloensis* would have had an independent origin. *F. daltoniana* might have contributed to the ancestry of *F. chiloensis*, since it has similar bright green leaves, although its current species distribution is far from the presumed northeast Asian origin of the octoploid species. It is also possible that the subspecies of *F. virginiana* with differing leaf coloration might have had independent origins. The glaucous leafed forms of *F. virginiana* might have had *F. iinumae* in their background, whereas the green-leafed subspecies did not.

The origin of Hawaiian and Chilean *F. chiloensis* is also obscure, but presumably they were introduced from North America via bird migrations (Darrow, 1931). Since all *F. chiloensis* in Hawaii are hermaphroditic, and those in Chile are gynodioecious, a northern introduction from Alaska or the Aleutian islands is most likely as hermaphrodites are more common there than in California. It is likely that multiple introductions were made into Chile as the habitats of South American *F. chiloensis* have an extensive range from beaches and headlands to montane forests at 1900 m elevation (Darrow, 1966; Cameron *et al.*, 1991, 1993). Glaucous forms of octoploids may also have been introduced, as at high elevations in Chile, leaf colour and thickness resembles *F. virginiana* ssp. *glauca* (Cameron *et al.*, 1993).

CONCLUSIONS

The direct progenitors of the dessert strawberry *F. × ananassa* are known to be *F. chiloensis* from South America and *F. virginiana* from the eastern US, but most other evolutionary relationships within the genus remain clouded. It is not even known precisely where the octoploids originated, and whether they had single or separate origins. It is likely that octoploids first appeared in the northeastern corner of Asia, perhaps in a form like *F. iturupensis*, but little field or experimental evidence is available to test this hypothesis.

There appears to have been only a limited amount of genomic divergence among the diploid species of *Fragaria*, making phylogenetic comparisons difficult without generating extensive molecular data. *F. vesca* may be the ancestral *Fragaria* species, as its range overlaps with most other species and its chromosomes will pair with many of them including the octoploids, but only limited numbers of hybridizations have been attempted between any species combination. Particularly limited is information on crossabilitites among European and Asian species. Until much more extensive studies are made among the lower ploidies, the evolutionary relationships within the group will remain subject to speculation.

Although numerous recent explorations have been made to collect North and South American octoploid strawberries for breeding purposes (see Chapter 4), little effort has been made to collect the Asian diploid species. Although the diploids are more difficult to use in breeding work, they deserve much more attention, if only because they hold the keys to better under-standing of *Fragaria* evolution. Extensive efforts need to be made to collect all the lower ploidies, and evaluate them both cytogenetically and with molecular markers.

3

HISTORY OF STRAWBERRY DOMESTICATION

It seems likely that our species, *Homo sapiens*, has always gathered and consumed strawberries from the wild. Who can walk by a strawberry patch in a forest or field without gathering these succulent berries? In fact, the ease with which strawberries can be collected from the wild, may actually have delayed their cultivation until almost modern times. Although our important grain crops were domesticated over 10,000 years ago (Hancock, 1992), the first strawberry species were domesticated in the last 2000 years, and the major strawberry of commerce, *F. × ananassa*, was born only 250 years ago.

DOMESTICATION OF OLD WORLD SPECIES

The strawberry was probably grown in Roman and Greek gardens, but there is only limited reference to its cultivation in early writings (Darrow, 1966). Ovid and Virgil mentioned the strawberry in poems, and Pliny (AD 23–79) lists its fruit 'Fraga' (fragrant) as a natural product of Italy. It seems likely that the Romans cultivated indigenous strawberries, as they spent considerable funds importing a wide array of fruits for their country estates including apples, apricots, cherries, citrons, figs, grapes, peaches, plums and pears (Wilhelm and Sagen, 1974).

The first references to strawberry cultivation in Europe appear in the French literature of the 1300s. Most notably, it is known that King Charles V had over 1000 strawberries planted in the royal gardens of the Louvre in Paris, and strawberries were grown in four blocks of the gardens of the Dukes of Burgundy (Darrow, 1966). The mother stock for these gardens was most likely collected from the wild, and then propagated by moving runners from established blocks to vacant soil. The popularity of the strawberry steadily grew during the Middle Ages, in spite of a warning from the noted abbess and mystic, St Hildegard von Bingen, in the 12th century that the strawberry was unhealthy because its fruit were found near the ground in stale air (Bühler, 1922).

It is known that *Fragaria vesca*, the wood strawberry or fraise des bois, was widely planted in gardens all across Europe by the 1500s. Clear records are common in Renaissance herbals, and after about 1530, there is a clear distinction made between wild and garden strawberries (Sauer, 1993). The wood strawberry was grown not only for private consumption, but for market as well. In fact, the strawberry may have got its name from the activities of street vendors who strung the berries on straws of grass or hay to take to market (Darrow, 1966; Wilhelm and Sagen, 1974). Another possible origin of the name relates to its ripening at the same time as hay, as *streaw* is the Anglo-Saxon word for hay. Most likely, strawberries were named after the way the runners 'strew' or scatter around the mother.

The first printed illustration of the strawberry is found in the *Herbarius Latinus Moguntiae* published in 1484 by Peter Schöffer, a partner of Johann Gutenberg (Leyel, 1926). The inclusion of the strawberry meant that it was considered important to healthful living. The first colour illustration of the strawberry (Plate 3) was published in 1485 by Schöffer in the German addition of his book called *Herbarius zu Teutsch* or *Gart der Gesundheit*, meaning literally 'Garden of Good Health'. This book 'occupied a place unsurpassed in German natural history for more than a half century' (Wilhelm and Sagen, 1974). In fact, the strawberry was grown widely in apothecary gardens all across Europe. All parts of the plants were used in medicinal teas, syrups, tinctures and ointments. Strawberry concoctions were used for skin irritations and bruises, bad breath, throat infections, kidney stones, broken bones and many other injuries.

Several different forms of *F. vesca* were identified by botanists in the 1500s, including albino types and everbearing ones from the Alps (*F. vesca semperflorens*). Some of the earliest cultivars were everbearing including 'Fraisier de Bargemont' from France, 'Haarbeer' and 'Brösling' ('Pressling') from Germany and 'Capiton' from Belgium. Most of these varieties were likely selected from the wild, except 'Capiton', which may have been derived from 'Haarbeer' (Wilhelm and Sagen, 1974). Fruit of all these varieties was pale coloured and only the early developing fruit had large size. Much redder Capitons with improved fruit size began to appear in the 16th century and they supplanted the white forms. Much like today, fruit from these early varieties were served with cream, soaked in wine or covered with powdered sugar. Strawberry jelly appeared in the 1600s (Wilhelm and Sagan, 1974).

The musky-flavoured *F. moschata* ('Hautbois' or 'Hautboy') was also planted in gardens by the late 15th century, together with the green strawberry, *F. viridis*. *Fragaria viridis* was used solely as an ornamental all across Europe, whereas *F. moschata* was utilized for its fruit by the English, Germans and Russians. The French largely scorned it (Duchesne, 1766). Domestication of the Hautboy probably began in the 16th century (Sauer, 1993) and the earliest cultivars, such as 'Fraisier à Bouquet', appeared in the 18th century. The origin of the musky strawberries was initially clouded by Philip Miller when he improperly suggested in his influential *Gardeners Dictionary* (1735) that they came from the New World, but in fact, they were of European origin (Wilhelm and Sagen, 1974). The name Hautboy was apparently an English spelling of the French *haut bois* (high [fruiting] woods [strawberry]).

Hautboy fruit varied from red to rose–violet and were borne on trusses extending above the leaves. Their flavour has been described as a mixture of honey and musk (Wilhelm and Sagen, 1974). The Hautboys initially fell into disrepute, as the earliest types were dioecious, leading to poor production when only one gender was planted, but the great French botanist Duchesne (1766) discovered that interplanting pistillate plants with good pollen producers of other *F. moschata* or *F. virginiana* yielded good crops. Regardless of this, perfect flowered types were soon developed.

By the 1600s, the culture of the native European strawberries was widely practised and well refined. Many of our modern practices had already been developed including frequent establishment of beds to maintain high plant vigour, the use of raised beds in areas with poor drainage, and the application of mulch to protect against winter cold. Fruit size was maximized by early planting dates, optimal plant spacing, elimination of the first flower trusses and removal of all but the first three or four flowers in a cluster. Europeans had become expert strawberry growers and the stage was set for the rapid acceptance of a new, highly promising horticultural type from the New World, *Fragaria virginiana*.

DOMESTICATION OF THE SCARLET STRAWBERRY, *FRAGARIA VIRGINIANA*

The wood strawberry, *F. vesca*, dominated strawberry cultivation in Europe for centuries, until *F. virginiana* from Canada and Virginia began to replace it in the 1600s. All of the clones that found their way to Europe were wild in origin, as the aboriginal peoples of North America did little gardening with strawberries. They enjoyed the fruit both fresh and in cornmeal bread (Wilhelm and Sagen, 1974; Sauer, 1993), but the natural abundance of the strawberry had generated little stimulus for domestication in the New World.

The exact particulars of the entity of *F. virginiana* into Europe are unknown, but it had certainly arrived in both France and England by the late 1500s, and new importations occurred regularly over the next 150 years. Jacques Cartier, the discoverer of the St Lawrence River of Canada in 1534, was probably the first to bring *F. virginiana* to the Old World. There are no specific records of its introduction by him, but Cartier made numerous mention of it in his diary and he was known to have introduced other Canadian plants (Wilhelm and Sagen, 1974). The first published references to the strawberry of Canada was in the garden catalogue of Robins, the botanist to Henri IV of France. He and his brother first reported it in 1624, but it is likely that they planted it long before then, as they were active importers of plants from throughout the world.

The Canada strawberry, then known primarily as *Fragaria americana* but now called *F. virginiana*, rapidly spread to gardens across France and all of Europe (Wilhelm and Sagen, 1974). It was incorporated in 1636 into the huge Jardin Royal des Herbes Mèdecinales of Guy de La Brosse, physician to Louis XIII. It appeared in a catalogue of Canadian exotics in 1633 written by Giovanni Battista Ferrari, a professor of Hebrew at the Collegio Romano in Rome. The noted apothecary, Jean Hermans, was growing the Canada strawberry in his garden in Brussels by 1652, and numerous English herbalists and horticulturalists were raising it by the early 1600s, including John Tradescant and John Parkinson. At least nine different accessions of *F. virginiana* were being grown in Europe by 1650.

Although the first strawberries imported from Canada were botanically interesting and intriguing horticultural curiosities, most produced little fruit, were green where the fruit were shaded and frequently produced excessive runners. Not until strawberries from Virginia became widely distributed did *F. virginiana* really make an impact on the horticultural industry. What became known as the Scarlet strawberry was favoured for its large fruit size, high yields and deep red colour. They were particularly enjoyed in jam, because of their persistent colour, acid flavour, high aroma and retention of shape (Darrow, 1966).

It is not clear when the first Scarlet strawberries arrived in Europe, but in the garden of de La Brosse he did have an item called *Fragaria americana magno fruto rubro*, which may have been a large-fruited, scarlet type. The surviving colonists from Plymouth may also have brought back Scarlet strawberries when they returned in 1586. Regardless, during the late 1500s and 1600s, all kinds of native plants of North America appeared in the gardens of Europe, making multiple introductions of the Scarlet strawberry likely.

The early cultivar development of *F. virginiana* was primarily conducted by growers who found raising seed imported from North America often resulted in horticulturally important variations. The numbers of varieties available increased dramatically from three to about 26, over a period of a few decades at the turn of the 17th century (Darrow, 1966). Some of the most important early cultivars were Oblong Scarlet, Grove End Scarlet, Duke of Kent's Scarlet, Knight's Large Scarlet, Wilmonts Late Scarlet, Morrisana Scarlet, Comman Scarlet and the Australian Scarlet and Hudson Bay Scarlet. They came from all over the New World including Nova Scotia, Virginia and New York. The early improvements were modest, however, and generally did not yield any substantial advancements from the best of the earlier imported types such as 'Large Early Scarlet'.

Garden culture of strawberries began in North America in the middle of the 17th century with varieties imported from England. Early garden calendars listed three types of strawberries: (i) the Hautboys (*F. moschata*); (ii) the Chili (*F. chiloensis* originally from Chile); and (iii) the Redwood (*F. vesca* from Europe) (Fletcher, 1917). Little cultivation of *F. virginiana* was undertaken until the importation of 'Large Early Scarlet' in the late 1700s, even though native strawberry populations abounded and 'Large Early Scarlet' had actually been sent to England a century earlier from the wilds of North America.

The first native American clone of *F. virginiana* to be propagated for sale in North America was called 'Hudson' (1791). It was very vigorous and had soft, scarlet fruit with high flavour. It was cultivated well into the 1800s and can probably be considered the first important American strawberry (Hedrick, 1925). The first commercial strawberry plantings were established around Boston, New York, Philadelphia and Baltimore in the early 1800s. Until that time, most commercial strawberries had been gathered from the extensive wild populations that were springing up in conjunction with the clearing of forests and the abandonment of worn-out agricultural sites.

'Red Wood', an English variety of *F. virginiana*, was probably the most important variety grown in the early part of the 18th century in North America, along with a number of Scarlet varieties (Fletcher, 1917). 'Red Wood' was thought to be inferior in flavour to the older Scarlet varieties, but had a longer season of production. Other popular cultivars during the first half of the 1800s were 'White Wood' (a white-fruited type), 'Early Hudson', 'Old Scarlet', 'Crimson Cone', 'Large Early Scarlet', 'Hudson's Bay', 'Methven Scarlet', 'Ross Phoenix', and 'Early Virginia'. All of these except for 'Ross Phoenix' and 'Hudson Bay' were imported from England.

DOMESTICATION OF *F. CHILOENSIS* IN THE NEW WORLD

The cultivated strawberry of South America, *Fragaria chiloensis*, has a long and rich history (Hancock *et al.*, 1999). It was utilized well over 1000 years ago by the indigenous Mapuches between the rivers Biobio and Tolten in south-central Chile, and the more northern Picunches tribe between the rivers Itata and Biobio. The Picunches had contact with the northern agrarian Inca invaders and were probably the first to transport elite plants from the wild to their home gardens. The Mapuches were primarily hunters and gatherers, but learned about agriculture from the Picunches.

Strawberry fruits were used by the native Chileans fresh, dried, as a fermented juice or as medicinal infusions against indigestion, diarrhoea and bleeding (de Moesbach, 1992). The Mapuches made many kinds of fermented juices, but their favourite was the one from the 'llahuen' or 'lahueñe' small red-fruited wild strawberry, that was called 'lahueñe mushca' (Labarca, 1994).

Most evidence indicates that the primary domesticants were the larger white-fruited forms, called 'kallén' or 'quellghen' by the Mapuches.

Albino-fruited types are rare in nature, but have been found at three southern locations. When subjected to a multivariate analysis, these forms were more closely associated with the cultivated white types than to the native wild red ones (Fig. 3.1).

Some red-fruited forms may also have been domesticated, but reports of their existence are sketchy. Darrow (1957a, b) described Chilean large red-fruited forms from around Santiago in the middle of the 20th century, but there are no earlier reports of cultivated 'lahuene'. Wild red-fruited forms were abundant from Santiago southward, so the pressure to cultivate them was probably minimal. It is reported, however, that the Mapuches planted small plots of the wild red forms in open spaces in the forests as a trap for the Spanish soldiers. When the soldiers dropped their arms to pick the fruit, the fierce Indians attacked and killed them (Gonzalez de Nájera, 1866).

Strawberry cultivation by the Mapuches was mostly limited to garden plots. After the conquest by the Spaniards, larger commercial plantings of 1–2 ha began to appear in the coastal areas from north of the Itata River to the Chiloé Island. These traditional plantings of *F. chiloensis* flourished until the 1950s, when they began to be mixed with northern hemisphere cultivars of

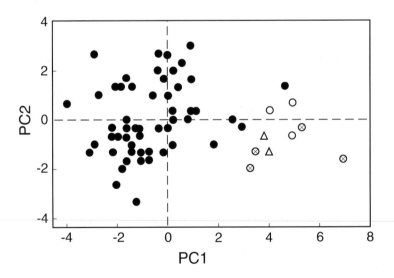

Fig. 3.1. Distribution of 61 Chilean accessions of strawberry on the first and second principal components (PC1 and PC2) of a multivariate analysis of morphological traits. Symbols represent accessions of: wild *Fragaria chiloensis* f. *patagonica* with red fruit (●); wild *F. chiloensis* f. *patagonica* with white fruit (△); cultivated *F. chiloensis* f. *chiloensis* (○), and cultivated *F. × ananassa* (⊗). Note that the white forms of the wild forms cluster closely with domesticated *F. chiloensis*. (Adapted from Lavín, 1997 and del Pozo, Muñoz, Lavín and Maureira, unpublished.)

F. × ananassa. F. chiloensis cultivation has now largely disappeared, but small plantings can still be found along the traditional area of cultivation from Iloca on the coast of Curicó province to the Chiloé Island (Hancock *et al.*, 1999).

SPREAD OF *F. CHILOENSIS* OUTSIDE CHILE

During their period of exploration and conquest in the mid to late 1500s, the Spanish spread *F. chiloensis* throughout northwestern South America. Major industries developed around Cuzco (Perú), Bogotá (Colombia) and Ambato (Ecuador) (Popenoe, 1921, 1926; Darrow, 1957a, b). The source of these plants is not known; however, the variability among the preserved land races suggests that they had multiple origins (Hancock *et al.*, 1999). The land races may have been spread from several Chilean locations or seedling volunteers may been moved from their original sites to new locations.

The largest hectarage of cultivated *F. chiloensis* in South America was grown at Guachi-Grande, Ecuador, near Ambato (Darrow, 1953). There were probably 500–700 ha from at least the late 1700s until 1970 (Hancock *et al.*, 1997; Finn *et al.*, 1998). Father Velasco wrote in 1789 that the 'frutilla' was three times the size of the European strawberry and 'it is produced throughout the entire year, and though it is common in several provinces, in no other is it so abundant, nor so excellent as in that of Ambato'. The English botanist Richard Spruce visited Ambato in the mid-1800s and proclaimed that the strawberry grown in abundance in the nearby village of Guachi was of exceptional quality. Wilson Popenoe (1921) declared that 'It is the custom in Ecuador to throw the fruits into boxes: they are then carried six or seven miles on mule-back to the city of Ambato, where they are sorted by hand, for shipment by train to Quito or Guayaquil. There is probably no other strawberry in the world which could tolerate this sort of handling.'

One of the Chilean clones even found its way into Europe in the 1700s compliments of a French spy, Captain Amédée Frézier (Darrow, 1966; Wilhelm and Sagen, 1974). Frézier was struck by the large-fruited strawberry grown around Concepción, Chile during his mapping of Spanish fortifications from 1714 to 1716. He selected some of the largest fruiting types and took them back with him to Marseilles in 1716 (Fig. 3.2). Five of these plants survived and one of them was given to Antoine de Jussieu, the director of the Jardin de Plantes in Paris, where clones of *F. virginiana* were already growing.

Reports on this remarkable new introduction spread widely, and within a few years, plants were located in botanical gardens all across Europe. Unfortunately, early reports on the Chilean strawberry were negative, as the plants were largely barren. Frézier had inadvertently brought back staminate plants that needed a pollinator. French horticulturalists solved the problem when they discovered that the 'Chili' would produce fruit when pollinated by *F. moschata* or *F. virginiana*.

Fig. 3.2. A woodcut of the original *Fraise du Chili* or *Frutilla* clone of Chilean *Fragaria chiloensis* brought to France by A. Frézier in 1716. The picture comes from Frézier's book *A Voyage to the South Seas and along the Coasts of Chile and Peru in the Years 1712, 1713 and 1714.* (From Darrow, 1966.)

The Chilean strawberry reached its highest acclaim in Brittany, where it came to be known as the Fraise de Plougastel, after one of the major cities of production. The Chili thrived in the cool maritime climate of Brittany, which was similar to its original home environment. By the mid-1800s, there was probably more *F. chiloensis* cultivated in France than in its native land, as 500 acres were grown in Brest and over 2500 acres in all of Brittany (Wilhelm and Sagen, 1974). The clones originally grown in Brittany had much more impressive size than the Scarlet types when effectively pollinated (primaries the size of walnuts), but their fruit were paler in colour (whitish red), more seedy, and fainter of flavour. Over time, improved hermaphroditic selections of *F. chiloensis* were identified with greater self-fertility, but they still needed a pollinator to reach maximal production.

The strawberry of Chile never became popular in Europe outside of

Brittany. Difficulties with its fertility probably played a role, but climatic factors may have been more important. The Chilean was difficult to grow in the harsher inland climates and had to be protected from winter cold. Under most continental conditions, the '. . . fruit was poor coloured, and poor textured and often had a mawkish flavour' (Darrow, 1966). Soil-borne pathogens were also a limiting factor, and the Chili grew vigorously only on well drained sandy soils (Wilhelm and Sagen, 1974).

ORIGIN OF THE DESSERT STRAWBERRY, *FRAGARIA × ANANASSA*

Unusual seedlings began to appear in Brittany and other gardens with unique combinations of fruit and morphological characteristics. Although the origin of these seedlings was initially clouded, Duchesne (Fig. 3.3) determined in 1766 that they were hybrids of *F. chiloensis × F. virginiana* and he named them

Fig. 3.3. The great French botanist, Antoine Nicholas Duchesne, who first recognized the hybrid nature of the garden strawberry, *Fragaria × ananassa*. (From Darrow, 1966.)

Fragaria × *ananassa* to denote the perfume of the fruit that smelled like pineapple (*Ananas*). It is not clear from the literature where the first hybrids of the pineapple or pine strawberry appeared, but they must have arisen early in the commercial fields of Brittany, and in botanical gardens all across Europe such as the Trianon, the Royal Garden at Versailles where Duchesne studied. The first hybrid cultivars were disseminated from the Netherlands, perhaps because the Dutch were such active seed merchants and had imported early hybrids, but it is also possible that they had recognized unique types in their own gardens. The first description of a variety that matched *F.* × *ananassa* was made by Philip Miller in the 1759 edition of the Gardener's dictionary, although he was not sure of its origin (Fig. 3.4).

Duchesne referred to these early cultivars of *F.* × *ananassa* as 'Quoimios' in his 1771 supplement to *L'Histoire Naturelle des Fraisiers* (Darrow, 1966). Two were pale fruited like the Chili, but were improved in other ways. One from Haarlem, the Netherlands was a partial hermaphrodite that was best used as a pollinator, as it bloomed during the same late season as the female Chili and its early developing flowers were barren. One called Quoimio de Bath had smaller berries than the Chilean, but was superior in vigour and size. Three other early varieties had much redder fruit than the Chili, including the 'Carolina' with cherry red fruit, Quimio de Cantorbéri which had deep coloured flesh and Clagny Quoimio, a scarlet coloured type, which Duchesne thought was a backcross of *F. virginiana* × Quoimio de Haarlem.

P L A T E   CCLXXXVIII.

FRAGARIA, *Tourn. Inst. R. H.* 295. *Tab.* 152. *Lin. Gen. Plant.* 558. Strawberry ; in French, *Fraisier*.

THIS Genus of Plants is ranged in the Seventh Section of *Tournefort's* Sixth Class, which includes the *Herbs with a Rose-flower, whose Pointal turns to a Fruit composed of many Seeds collected into a Head.* Dr. *Linnæus* ranges it in the Fifth Section of his Twelfth Class, which includes those Plants whose Flowers have about Twenty Stamina, which are inserted in the Empalement, and many Styles.

The Characters of the Genus are,

The Empalement of the Flower is plain, of One Leaf; which is cut deeply into Ten or Twelve acute Segments, which are alternately large and small, and is permanent. The Flower has Five or Six roundish Petals, which are inserted in the Empalement; and about Twenty Awl-shaped Stamina, which are shorter than the Petals, terminated by lunular Summits; and a great Number of small Germina collected into a roundish Head, each having a Style inserted to their Sides, crowned by single Stigmas. The Germina afterward turn to an oval pulpose soft Fruit sitting on the Empalement, having many small Seeds.

The Species here represented is,

FRAGARIA *foliis ovatis crenatis nervosis, calycibus maximis.* Strawberry with oval crenated veined Leaves, and the largest Empalement.

This Sort of Strawberry has been of late Years introduced into the *English* Gardens; but from whence it came we are at a loss to know. Some Persons have affirmed it was brought from *Louisiana*; others that it came from *Virginia*; but I recieved some Plants of this Kind from a curious Gentleman of *Amsterdam*, who assured me they were brought from *Surinam*.

The Leaves of this Sort have a great Resemblance to those of the Scarlet Strawberry; but are larger, of a thicker Substance, and the Indentures of their Edges are blunter : The Runners from the Sides of the Plants are much larger, and are hairy : The Footstalks of the Flowers are stronger, the Flowers are much larger, and the Fruit approaches in Size, Shape, and Colour, to the *Chili* Strawberry. Whether this is a distinct Species, or an accidental Variety which came from Seeds, I shall not take upon me to determine, since it has an Affinity to Two or Three other Species. But as this Sort produces a great Quantity of Fruit, when the Plants are kept clear from Runners, and the Fruit is very large, so it is worthy of Cultivation.

a represents the Empalement of the Fruit, which is deeply cut into Twelve acute Segments, Six of which are alternately larger than the other. *b* shews the Six round Petals of the Flower; *c* the numerous Stamina and Styles in the Center; and *d* represents the Fruit, when ripe, of the natural Size. *e* shews the Empalement separated from the Fruit; *f* One of the Stamina, with its Summit; *g* One of the Styles, sitting on the Germen, and *h* One of the Seeds separated from the Fruit.

Fig. 3.4. The first European description of *Fragaria* × *ananassa* made by Philip Miller in the 1759 edition of his *Gardener's Dictionary*. (From Otterbacher and Skirvin, 1978.)

BREEDING IN EUROPE FROM 1800 TO 1960

At first, only chance hybrids of *F. chiloensis* and *F. virginiana* were evaluated by horticulturalists. Steady genetic progress was made over the years, but yields and fruit quality still left something to be desired. Formal strawberry breeding was initiated in England in 1817 by Thomas A. Knight (Pearl, 1928; Darrow, 1966; Wilhelm and Sagen, 1974). He was one of the first systematic breeders of any crop, and used clones of both *F. virginiana* and *F. chiloensis* in his crosses. He produced the famous 'Downton' and 'Elton' cultivars, noted for their large fruit, vigour and hardiness. Michael Keen, a market gardener near London, also became interested in strawberry improvement about this time and developed 'Keen's Imperial' whose offspring, 'Keen's Seedling' (Fig. 3.5), is in the background of many modern cultivars. The berries of 'Keen's Seedling' were a sensation as they were sometimes 2 inches in diameter, deep red in

Fig. 3.5. The strawberry 'Keen's Seedling', which was a sensation in England in the late 1800's. It is in the pedigree of many modern cultivars.

colour, and particularly good flavoured. The plants were prolific and bore their fruit well above the ground. This variety dominated strawberry acreage for close to a century (Table 3.1) and The Royal Horticultural Society awarded this remarkable berry its silver cup in 1821.

Table 3.1. Dominant *Fragaria* × *ananassa* cultivars in Europe before about 1975. (Sources: Brooks and Ohmo 1996; Darrow, 1966.)

Cultivar	Year of release	Place developed	Notable characteristics
Downton	1817	England	Large, oblong fruit
Elton	1828	England	Lateness, beautiful fruit, vigour and hardiness
Keen's Seedling	1821	England	Very large fruit, deep rich red colour, juicy and vigorous with upright fruit
British Queen	1840	England	Excellent flavour
Héricart de Thury	1849	France	Excellent flavour, glossy red colour and firmness
Jucunda	1854	England	Productivity, firmness, red flesh colour, lateness and easy capping ability
Marguerite	1859	France	Huge fruit size and acid flesh
Sir Joseph Paxton	1862	England	Brilliant, firm and glossy red fruit
Nobel	1884	England	Large, attractive fruit, productivity and broad adaptation
Royal Sovereign	1892	England	Large, bright scarlet, firm and high flavoured fruit
Deutsch Evern	1902	Germany	Early, long conic fruit
Madame Moutot	1910	France	Productivity, very large fruit and broad adaptation
Oberschlesien	1919	Germany	Productivity and broad adaptation
Surprise des Halles	1910	France	Earliness and productivity
Auchincruive Climax	1947	Scotland	Productivity and high fruit quality
Cambridge Favourite	1947	England	Long picking season, excellent dessert and canning quality
Cambridge Vigour	1947	England	Large, firm fruit and drought tolerance
Senga Sengana	1954	Germany	Hardiness, broad adaptation, productivity, deep red colour and excellent processing quality
Talisman	1955	Scotland	Similar to Auchincruive Climax, but highly resistant to red stele root rot
Redgauntlet	1956	Scotland	Maintenance of size throughout season, bruise resistance and resistant to red stele
Gorella	1960	Netherlands	Heathy foliage and very large

Numerous excellent varieties followed the success of 'Keen's Seedling' in Europe in the first half of the 18th century and were grown for decades (Darrow, 1966). Among the most elite were J. William's 'Pitmaston Black', a Mr Myatt's 'British Queen' (1840), J. Salter's 'Jucunda' (1854), and J. Bradley's 'Sir Joseph Paxton' (1862). 'Pitmaston Black' and 'British Queen' were dominant in the mid-1800s and were grown in England as late as 1914. 'Jucunta' was a major variety in both Europe and the USA until the 1920s, and was still planted to some extent in the 1960s. 'Sir Joseph Paxton' was the leading variety in England until the 1930s and in France for another decade. Particularly noteworthy characteristics of these cultivars were the high flavour of the 'Pitmaston Black' and 'British Queen', the productivity, lateness, high flavour and capping ease of 'Jucunda', and the brilliant, glossy red colour and firmness of 'Sir Joseph Paxton'.

Several excellent French varieties were released during the mid-1800s including J. Jamin's 'Héricart de Thury' (1845) and M. Lebreton's 'Marguerite' (1858) (Darrow, 1966). 'Héricart de Thury' had excellent flavour and glossy red, firm berries that made it the most important preserve berry for 100 years. 'Marguerite' did not achieve the long-term status of 'Héricart de Thury', but was widely admired for its exceptionally large berries, said to average 15–20 g with 40–45 g primaries.

Thomas Laxton of England was the most active breeder during the later part of the 18th century and released a number of important varieties including 'Noble' (1884) and 'Royal Sovereign' (1892). These two varieties were grown on both sides of the Atlantic Ocean, and were popular until the middle of the 20th century. 'Noble' was known for its earliness, cold hardiness and disease resistance. 'Royal Sovereign' was popular because of its earliness, productivity, flavour, attractiveness, and hardiness. The American variety 'Sharpless' was in the ancestry of both these cultivars.

Some of the most popular releases in the early 20th century were: C. Moutot's 'Madame Moutot' (France, 1910), Bottner's 'Deutsch Evern' (Germany, 1902), O. Schindler's 'Oberschlesien' (Germany, 1919) and Guyot of Dijon's 'Surprise des Halles' (France, 1929).

'Madame Moutot' was popular in France and other European countries until the late 1960s due to its size and productiveness. 'Deusch Evern' was the standard early variety in northern Europe for decades and was still grown to some extent in 1960 (Darrow, 1966). It was noted for its productivity and long, conical, light-red fruit. 'Oberschlesien' was widely planted in Germany until 1945 for its high yield and adaptability. 'Surprise des Halles' was the most popular cultivar in France in the mid-1960s because of its earliness, fruit quality and productivity.

In the middle of the 20th century, a number of particularly active breeding programmes emerged in Scotland, England, Germany, and The Netherlands. In Scotland, R. Reid developed a series of red stele resistant varieties utilizing American 'Aberdeen' as a source of resistance. His variety

'Auchincruive Climax' (1947) dominated acreage in Great Britain and northern Europe until its demise due to June yellows in the mid 1950s. He then released 'Redgauntlet' (1956) and 'Talisman' (1955), which served as suitable replacements. In England, D. Boyle produced a large series of varieties with the prefix 'Cambridge'. 'Cambridge Favourite' (1953) became the most important of the group and dominated the acreage in Great Britain by the 1960s. It is still planted somewhat today, due to its productivity, firmness, ship ability and capping ease. In Germany, R. von Sengbusch's produced a 'Senga' series, of which 'Senga Sengana' (1954) became paramount. 'Senga Sengana' was widely planted for its processing quality and is still important in Poland and other eastern European countries. In the Netherlands, H. Kronenberg and L. Wassenaar released several cultivars, of which 'Gorella' (1960) made the greatest impact. It was noted for its size, bright red glossy skin and red flesh.

BREEDING OUTSIDE EUROPE FROM 1800 TO 1960

The greatest concentration of breeding activity outside of Europe until the modern period was in the US, although the Japanese produced two important varieties: Dr H. Fukuba's 'Fukuba' (1899), noted for its large size and high flavour (Darrow, 1966), and K. Tamari's 'Kogyoku' (1940), respected for its vigour, earliness and fruit size (Mochizuki, 1995). 'Fukuba' was the most important variety in forcing culture until the early 1970s. 'Kogyoku' was one of the leading field grown cultivars after World War II, until it lost importance to the American import 'Donner' in the 1950s (Darrow, 1966).

Charles Hovey, of Cambridge, Massachusetts, produced the first important North American cultivar, 'Hovey', by crossing the European pine strawberry, 'Mulberry' with a native clone of *F. virginiana* in 1836 (Plate 4). It was the first variety of any fruit to come from an artificial cross in America and for some time made the strawberry the major pomological product in the country (Hedrick, 1925; Table 3.2).

'Hovey' often proved to have low fruit set and the basis of this led to considerable horticultural debate (Fletcher, 1917). It had been known in Europe since 1760 that strawberries could have separate genders, but Americans were not aware of this work and for a decade from 1840 to 1850, debate raged among horticultural societies as to the nature of 'Hovey's unproductiveness. Hovey himself, originally asserted that his cultivar was perfect, but eventually had to recant under pressure from Nicholas Longworth of Cincinnati, Ohio.

The release of 'Hovey' stimulated a great deal of interest in strawberries throughout the country and numerous private individuals began making crosses and growing seedlings. The variety that had the largest immediate impact was 'Wilson' (1851), developed by James Wilson in New York. It played

Table 3.2. Dominant *Fragaria* × *ananassa* cultivars in North America before 1975.

Cultivar	Year of release	Place developed	Notable characteristics[a]
Hovey	1836	Massachusetts	First variety of fruit developed in North America; better adapted than English varieties
Wilson	1851	New York	Productivity, firmness, deep red colour, hardiness and broad adaptability
Sharpless	1872	Pennsylvania	Pollinator of 'Wilson'
Crescent	1876	Connecticut	Productivity, earliness, hardiness and broad adaptability
Ettersburg 80	1880	California	Drought resistant, dessert and jam quality, hardiness
Marshall	1890	Massachusetts	Adaptation to mild climates; large fruit size, excellent flavour, high freezing quality and drought tolerance
Nich Ohmer	1898	Ohio	Productivity, large, firm, glossy-crimson berries
Aroma	1889	Kansas	Productivity; large, firm, attractive fruit, well adapted to clay and silt soils, resistance to leaf spot and leaf scorch
Pan American	1898	New York	First widely successful everbearer
Missionary	1900	Virginia	Low chilling requirement
Dunlap	1900	Illinois	Extreme hardiness, high flavour and deep red colour
Klondike	1901	Louisiana	Tolerance to heat, deep red colour and vigour
Howard 17 (Premier)	1915	Massachusetts	Excellent parent; resistance to leaf diseases, early flower bud initiation, multiple crowns, deep red colour and productivity
Blakemore	1929	Maryland	Outstanding shipper, resistant to leaf diseases, scarlet colour and productivity
Sparkle	1931	New Jersey	Productivity; glossy, attractive fruit, fresh and frozen quality
Robinson	1932	Michigan	Large size; productivity; colour and tolerance to virus diseases
Fairfax	1933	Maryland	Outstanding flavour, resistance to leaf diseases, low acid fruit
Catskill	1933	New York	Large fruit size and productivity
Gem	1933	Michigan	Dependable everbearer
Tennessee Beauty	1933	Tennessee	Productivity, firm and glossy red fruit, tolerance to virus diseases, capping ease

Table 3.2. *Continued*

Cultivar	Year of release	Place developed	Notable characteristics[a]
Sparkle	1943	New Jersey	Productivity, attractiveness, high flavour and resistance to red stele
Albritton	1945	N. Carolina	Late, large, uniform, attractive and firm fruit with excellent flavour, freezing quality
Lassen	1945	California	Low chilling requirement, large fruit and productivity
Shasta	1945	California	Large, firm and attractive berries
Pocahontas	1946	Maryland	Large, firm and attractive fruit, good frozen quality
Northwest	1949	Washington	Lateness, productivity and tolerance to virus
Surecrop	1950	Maryland	Resistance to many diseases (red stele, leaf spot, leaf scorch and verticillium wilt), vigorous and drought resistant
Midway	1951	Maryland	Productivity, deep red colour, freezing quality and resistance to red stele
Tioga	1955	California	Large, attractive, firm fruit, high productivity
Headliner	1957	Louisiana	Early ripening, large fruit, productivity and resistance to leaf spot
Redcoat	1957	Ontario	High yield, appearance, earliness and shipping qualities
Goldsmith	1958	California	Productivity, large, firm, glossy, attractive berries, good shipping quality
Florida 90	1952	Florida	Long, large and early berries; high flavour and productivity
Daybreak	1961	Louisiana	High productivity , large and attractive fruit, excellent flavour, good shipping quality

[a]Sources: Darrow (1966) and Brooks and Olms (1997).

an important role in expanding the North American strawberry industry from a few thousand acres to hundreds of thousands. It produced large, dependable crops even under indifferent care and its flowers were bisexual, eliminating the need for a pollinator. 'Wilson's' fruit were larger and more attractive than its predecessors, and its fruit were firm enough to ship long distances. Unfortunately, its flavour was very sour, but its other strengths led to the liberal use of the sugar bowl.

Several important cultivars were developed in the late 1800s and early

1900s to meet the need of the burgeoning strawberry industry (Darrow, 1937, 1966). Some of the most successful were: W. Parmalee's 'Crescent' (Conneticut, 1876), J. Sharpless's 'Sharpless' (Pennsylvania, 1872), Neunan's 'Neunan' (South Carolina, 1868), M. Ewell's 'Marshall' (Massachusetts, 1890), C. Loftus's 'Banner' (California, about 1890), E. Cruse's 'Aroma' (Kansas, 1892), J. Beaver's 'Nich Ohmer' (Ohio, 1898) and S. Cooper's 'Pan American' (New York, 1898). 'Neunan' was a seedling of 'Wilson' and became the standard in the southeastern US in the late 1800s, because it was a little less dark and a touch firmer than 'Wilson' in hot climates. 'Crescent' and 'Sharpless' were second and third to 'Wilson' in popularity from 1880 to 1900. 'Wilson' was pistillate and 'Sharpless' was commonly used as its pollinator. 'Marshall' was only a modest success in the east where it was bred, but became a major variety in the Pacific Northwest and California from 1905 to almost 1960, because of its high flavour and how well it could be frozen and preserved. 'Pan American' was notable as the first really successful ever-bearing variety. Although it was widely planted in gardens, its greatest importance was as a parent and was a major source of the everbearing trait until Powers and Bringhurst utilized native populations of *F. virginiana* ssp. *glauca* (see Chapter 4). 'Nich Ohmer' was not successful in its state of origin, Ohio, due to its low vigour, susceptibility to leaf spot and its small, only fair-flavoured fruit, but it was widely grown in California in the 1920s and 1930s because of its long fruiting season. It is in the genetic background of most successful California cultivars.

'Banner' ('Sweet Briar') was discovered in an abandoned strawberry patch by C. Loftus on his farm in Sweet Briar, California (Wilhelm and Sagen, 1972). Its early importance was limited due to insufficient runner production, but it rapidly grew in prominence in the central Valley from 1904 until the mid-1930s, when it was found that it could be successfully propagated in northern California. The berries of the 'Banner' strawberry were large, regularly conical in shape, bright red and had a wonderful aroma and fragrance. It may have been a seedling or runner descendant of 'Marshall'.

Some of the most important varieties developed in the early part of the present century were (Darrow, 1966): Rev. J. Reasoner's 'Dunlap' (Illinois, 1900), N. Gohn's 'Missionary' (Virginia, 1900), R. Cloud's 'Klondike' (Louisana, 1901), A. and E. Howard's 'Howard 17' or 'Premier' (Massachusetts, 1915) and J. Kuhn's 'Aberdeen' (New Jersey, 1923). 'Dunlap' dominated acreage in the northern states and Canada for the first 50 years of the century, because of its extreme hardiness. It was used widely as a parent in everbearing breeding, even though it was a short-day type. 'Missionary' was also import-ant for about 50 years, as one of the first really good low chilling varieties, performing well even in the semitropics of Florida. It was also an excellent parent in breeding of warm climate types. 'Howard 17' ('Premier') was a major eastern cultivar for at least 40 years both as a commercial cultivar and breeding parent. It is found in the ancestry of most North American and many

European cultivars, and was particularly noted for its resistance to leaf diseases and viruses, formation of many crowns, frost-hardy flowers and high productivity. 'Aberdeen' was grown in the 1930s in New Jersey and the east, but may have found its greatest importance as a breeding source for red stele resistance in both European and North American cultivars.

The most active breeder around the turn of the century was probably Albert Etter of California who developed dozens of varieties utilizing native *F. chiloensis* clones (Wilhelm and Sagen, 1974; Fishman, 1987). His most successful variety was 'Ettersburg 80' (1910), which was widely grown in California, Europe, New Zealand and Australia. Renamed as 'Huxley', it was still popular in England as late as 1953. 'Ettersburg 80' was extremely drought resistant, of outstanding dessert and jam quality due to its solid bright red colour, and was unusually hardy for a California type. Other outstanding Etter varieties were 'Ettersburg 121', 'Fendalcino' and 'Rose Ettersburg'. Although his releases were very successful as cultivars, they may have had their greatest impact as breeding parents. Hardly any California cultivar (and many others) do not have an Ettersburg variety in their background (Darrow, 1937, 1966; Sjulin and Dale, 1987).

In the 1930s and early 1940s, several new cultivars became important across the US including the great USDA breeder G. Darrow's 'Blakemore' (USDA-MD, 1929) and 'Fairfax' (USDA-MD/NC, 1933), G. Slate's 'Catskill' (New York, 1933), K. Keplinger's 'Gem' (Michigan, 1933), J. Haley's 'Robinson' (Michigan, 1940) and E. Henry's 'Tennessee Beauty' (Tennessee, 1943). 'Blakemore' became the major southern variety in the mid-1930s until the early 1960s because of its earliness, firm, bright red berries, suitability for freezing and preserving, and disease resistance. 'Blakemore' was used extensively in breeding, finding its way into the ancestry of a diverse array of cultivars grown in all parts of the US. 'Tennesee Beauty' became dominant in the upper south from 1940 to the 1960s due to its high productivity, tolerance to virus and good freezing quality. 'Gem' (also called 'Superperfection' and 'Brilliant') was the most important everbearer from 1940 to the modern period. 'Fairfax' was widely planted in the middle of this century from southern New England to Maryland and westward to Kansas. It was particularly noted for its outstanding flavour, but may have been more important as a breeding parent, finding its way into the pedigree of numerous European, US, Canadian and even Japanese cultivars.

In the late 1940s and early 1950s, several more cultivars achieved importance including J. Clarke's 'Sparkle' or 'Paymaster' (New Jersey, 1943), E. Morrow's 'Albritton' (North Carolina, 1945), H. Thomas and E. Goldsmith's 'Lassen' and 'Shasta' (California, 1945), D. Scott and G. Darrow's 'Pocahontas' (Maryland, 1946), and C. Schwartze's 'Northwest' (Washington, 1949). 'Sparkle' dominated in the northeast and midwest in the 1950 and 1960s, due to its high flavour, attractive appearance and resistance to red stele. 'Northwest' was the most planted variety in the US in the 1960s, even though all of the acreage was in Oregon and Washington (Darrow, 1966). It was particularly noted for its lateness and tolerance to virus diseases. 'Shasta' was widely grown in the central coast of California in the 1950s and 1960s because of its large size, firmness and long season. 'Lassen', grown extensively in southern California about the same period, was prized for its short rest period and high productivity. 'Pocahontas' was widely grown in the lower midwest and south in the 1960s and 1970s, and even became important in Italy in the late 1970s. It was known for its productivity, large, attractive, firm berries and good freezing quality.

The middle decades of the 20th century saw the release of D. Scott's 'Surecrop' (Maryland, 1950) and 'Midway' (Maryland, 1960), R. Bringhurst and V. Voth's 'Tioga' (California, 1955), P. Hawthorne's 'Headliner' (Louisiana, 1957) and 'Dabreak' (Louisiana, 1961), L. Spangelo's 'Redcoat' (Ontario, 1957), H. Thomas and E. Goldsmith's 'Goldsmith' (California, 1958), and A. Brook's 'Florida 90' (Florida, 1952). 'Surecrop', which was important in the upper south and lower midwest during the 1960s and 1970s, was noted for its high disease tolerance to both to leaf and soil pests. 'Midway' replaced Robinson in the 1960s in the midwestern US, primarily because of its firmer berries and resistance to red stele. 'Redcoat' dominated eastern Canada in the 1960s and 1970s. It was known for its high yield, appearance, earliness and shipping qualities. 'Headliner' and subsequently 'Dabreak' became important in the south in the 1960s and 1970s, replacing the earlier varieties, due to their larger size, higher productivity and an earlier season. 'Tioga' replaced 'Lassen' in California in the late 1960s because of its greater size, attractiveness, firmness and productivity. 'Florida 90' became popular in Florida late 1950s due to its very long, large, early berries and high flavour (Darrow, 1966). 'Goldsmith' was the first important private variety in California, noted for its productivity and shipping quality.

CONCLUSIONS

Strawberries were probably domesticated in the last two millenia. There is written evidence that *F. vesca* was planted in early Greek and Roman gardens, and we know that *F. chiloensis* was cultivated in South America by the native Chilenos over 1000 years ago. By the Middle Ages, *F. vesca* was planted widely across Europe, and *F. moschata* was a popular garden plant in France. The Spanish conquistadors considered the strawberry to be an important spoil of war and during the second half of the 15th century, spread *F. chiloensis* all across northwestern South America.

Clones of *F. virginiana* began to arrive in Europe from North America by the late 1500s, followed over a century later by *F. chiloensis* from South America. *F. virginiana* rapidly became an important garden plant all across Europe, but the importance of *F. chiloensis* remained much more modest until the two species accidentally hybridized in the late 17th century, probably in France. The *F. chiloensis* plants that had been brought to Europe were female and had insufficient hardiness to be widely grown. The resulting hybrid, *F. × ananassa*, combined the hardiness, vigour and productivity of *F. virginiana* with the large fruit size of *F. chiloensis*.

It did not take very long until breeders began systematically to improve *F. × ananassa*. 'Keen's Seedling' (1819) was the first major success in England, whereas 'Hovey' (1836) was the first really important American cultivar. Breeding activity on both sides of the Atlantic remained high in the 18th and 19th centuries, with a regular flow of varieties being produced. Initially, successful varieties such as 'Jucunda', 'Sir Joseph Paxton', 'Héricart de Thury' and 'Royal Sovereign' remained popular across large areas for many decades, but as the 20th century progressed, cultivar distribution became more regional and short-lived, particularly in the US where distinct assemblages of cultivars appeared in the eastern, central, southern and western parts of the country. As documented in Chapter 1, this regionalization of cultivars largely remains today, with the exception of the California-bred varieties, which now dominate all Mediterranean climates.

4

BREEDING AND GENETICS

INTRODUCTION

As outlined in Chapter 3, the dessert strawberry was one of the first crops to be systematically bred, beginning with the pioneering work of Thomas A. Knight in the early 1800s. Over the last two centuries, a constant stream of cultivars has been generated in Europe and North America, and more recently all across the world.

Much has been learned about the genetics of the strawberry since Duchesne led the way with his careful studies of sex and interspecific hybridization in the mid-1700s. The genus *Fragaria* is a particularly interesting subject of study to botanists and breeders alike, as several different reproductive strategies and ploidies are represented across a particularly broad eco-geographic range. The strawberry is quite amenable to research study, as the presence of stolons makes designing replicated experiments relatively easy.

The hybrid nature of the domesticated strawberry, *F. × ananassa*, provides considerable opportunity to utilize wild germplasm in breeding. The progenitors of the dessert strawberry, *F. chiloenis* and *F. virginiana*, are completely interfertile and have extensive natural ranges. This has stimulated numerous ecological studies that complement breeding efforts. In this chapter, the methods most commonly used in strawberry improvement are first described, together with the genetics of the most important horticultural traits. Germplasm resources are then reviewed in light of varietal improvement.

BREEDING STRATEGIES

The dessert strawberry is an out-crossed crop that is relatively sensitive to inbreeding (Morrow and Darrow, 1952; Melville *et al.*, 1980b), and as a result, most varietal improvement programmes have been based on pedigree breeding where elite parents are selected each generation for intercrossing. If adequate population sizes are maintained, changes in levels of homozygosity across generations appear to be minimal (Shaw, 1995).

Since highly heterozygous genotypes can be tested and maintained as runners, it seems there is little advantage to producing inbreds; however, selfing has been successfully employed in a number of instances to concentrate genes of interest. Anstey and Wilcox (1950) and Jones and Singleton (1940) produced superior seedling populations by combing second- and third-generation inbred lines. Craig *et al.* (1963) used recurrent reciprocal selection to reduce variability among seedlings and raise the general level of the characteristics. Spangelo *et al.* (1971a) found that S_5 inbreds of 'Howard 17' produced superior progenies to 'Howard 17' itself. Craig *et al.* (1963) observed that selfing significantly improved plant stand, foliage health, foliage uniformity, berry uniformity and berry appearance. In fact, several cultivars have been developed from selfs including 'Albritton' (Morrow and Darrow, 1952), 'Reiko' (Mochizuki, 1995) and 'Aliso' (Hancock *et al.*, 1996).

Backcrossing has also been used in a few instances to incorporate specific traits. Barritt and Shanks (1980) moved resistance to the strawberry aphid from native *F. chiloensis* to *F. × ananassa*. Bringhurst and Voth (1978, 1984) transferred the day-neutrality trait from native *F. virginiana* spp. *glauca* to *F. × ananassa*. Approximately three generations were necessary to restore fruit size and yield to commercial levels.

GENETICS OF ECONOMICALLY IMPORTANT TRAITS

Numerous comprehensive studies have been made on the genetics of the commercially important quality traits and yield of strawberries. Most of these traits have proved to be quantitative and highly heritable, but actual values often vary greatly among studies. This is not surprising as estimates of genetic variation are always closely linked with individual breeding populations and test environments (Galletta and Maas, 1990; Shaw, 1991).

Pest resistance

There have been a number of studies on the genetics of resistance to diseases and pests (Table 4.1). Resistant genotypes have been identified for alternaria leaf spot (black leaf spot) (*Alternaria alternaria* (FR.) Keissler), ramularia leaf spot, (*Mycosphaerella fragariae* (Tul.) Lindau), leaf scorch (*Diplocarpon earliana* (Ell. & Everh.) Wolf), powdery mildew (*Sphaerotheca macularis* (Fr.) Magn.), verticillium wilt (*Verticillium albo-atrum* Reinke & Berth. and *V. dahlia*), red stele root rot (*Phytophthora fragariae* Hickman), anthracnose (*Colletotrichum fragariae* Brooks), cactorum crown rot (*Phytophthora cactorum* (Leb. & Cohn) Schroet), two-spotted spider mite, virus tolerance, and berry mould resistance.

Table 4.1. Inheritance patterns of disease and pest resistance in strawberries.

Disease or pest	Observations	Representative studies
Fungal diseases of leaves		
Alternaria leaf spot (black leaf spot) — *Alternaria alternaria*	Resistant types were identified; resistance appears to be region specific; susceptibility is inherited as a single locus expressing dominance	Cho and Moon (1980), Yamamoto *et al.* (1985), Wassenaar and van der Scheer (1989), Takahashi *et al.* (1990)
Anthracnose — *Colletotrichum fragariae*	Resistant types were identified; no single genotype is resistant to all pathotypes; types resistant to *C. fragariae* tend to be resistant to *C. acutatum*	Delp and Milholland (1981), Smith and Black (1987), Galletta *et al.* (1993)
Anthracnose — *Colletotrichum acutatum*	Resistant types were identified; appears to be major gene action involved in resistance	Denoyes-Rothan (1997), Galletta, *et al.* (1993), Gupton and Smith (1991), Simpson *et al.* (1994)
Leaf scorch — *Diplocarpon earliana*	Resistant types were identified; the resistance is environmental and biotype specific; parents are not useful in predicting resistance of progeny	Nemec and Blake (1971)
Leaf spot — *Mycosphaerella fragariae*	Resistant types were identified; moderate levels of heritability	Sato *et al.* (1965), Nemec (1971), Shaw *et al.* (1988), Delhomez *et al.* (1995),
Powdery mildew — *Sphaerotheca macularis*	Resistant types were identified; general and specific combining ability is important; cuticle thickness may relate to resistance	Darrow *et al.* (1954), Daubeny (1961), Hsu *et al.* (1969), Oydvin (1980), Simpson (1988)
Fungal diseases of fruit		
Grey mold — *Botrytis cinerea*	Resistant genotypes were identified; resistance appears to be quantitative and additive; fruit firmness may be related to resistance	Daubeny and Pepin (1977), Maas and Smith (1978), Barritt (1980), Popova *et al.* (1985)
Fungal diseases of root and crown		
Black root rot — non-specific (several species)	Resistant genotypes were identified	Wing *et al.* (1995)

Table 4.1. *Continued*

Disease or pest	Observations	Representative studies
Cactorum crown rot — *Phytophthora cactorum*	Resistant octoploid genotypes were identified, *F. vesca* may be a better source of resistance	van der Scheer (1973), Seemuller (1977), Gooding *et al.* (1981), Bell *et al.* (1997), Rijbroek *et al.* (1997)
Fusarium wilt — *Fusarium oxysporium*	Resistant genotypes were identified	Honda *et al.* (1981), Kim *et al.* (1982), Cho and Moon (1984)
Red stele rootrot — *Phytophthora fragariae*	Resistant genotypes were identified; inbreeding can concentrate resistance genes; specific combining ability is very important; data fits a gene for gene model	Daubeny (1963), Montgomerie (1967), Melville *et al.* (1980a), van de Weg (1997)
Verticillium wilt — *Verticillium albo-atrum* and *V. dahlia*	Resistant genotypes were identified; additive variation is important; resistance is partially dominant, but breeding success relies on recovery of transgenic segregants	Wilhelm (1955), Gooding (1972), Arulsekar (1979), Maas *et al.* (1989), Shaw *et al.* (1996)
Virus and phytoplasma diseases		
Arabis mosaic	Little resistance was identified	Murant and Lister (1987)
Clover phyllody	Resistant genotypes were identified	Chiykowski and Craig (1975)
Raspberry ringspot	Little resistance was identified	Murant and Lister (1987)
Strawberry latent ringspot	Little resistance was identified	Murant and Lister (1987)
Tomato black ring	Resistant genotypes were identified	Murant and Lister (1987)
Tomato ringspot	Resistant genotypes were identified	Converse (1987)
Miscellaneous virus complexes	Resistant genotypes were identified to regional virus complexes including yellows; resistance in an area is related to the virus strains present and aphid pressure; a high proportion of the genetic variability is additive	Barritt and Daubeny (1982), Graichen *et al.* (1985), Sjulin *et al.* (1986)
Blossom and fruit pests		
Blossom weevil —	Resistance is under	Simpson *et al.* (1997)

Table 4.1. *Continued*

Disease or pest	Observations	Representative studies
Anthonomus rubi	independent genetic control from flowering time	
Tarnished plant bug – *Lygus lineolaris*	Resistant genotypes were identified	Tingey (1976), Schaefers (1980), Handley *et al.* (1991)
Leaf pests		
Strawberry aphid – *Chaetosiphon fragaefolii*	Resistant genotypes were identified; is regulated by more than one locus, with partial dominance and additive action, but highly resistant types are recoverable in backcross generations	Barritt and Shanks (1980), Crock *et al.* (1982), Shanks and Barritt (1974)
Cyclamen mite – *Steneotassonemus pallidus*	Resistant genotypes were identified	Breakey and Dailey (1956), Oydvin (1980)
Two-spotted spider mites – *Tetranychus urticae*	Resistant genotypes were identified; resistance may be biotype specific; strong additive and dominance effects involved in resistance	Schuster *et al.* (1980), Barritt and Shanks (1981), Crock *et al.* (1982), Shanks *et al.* (1995)
Root and crown pests		
Black vine weevil – *Otiorhynchus sulcatus*	Tolerant genotypes were identified; probable quantitative inheritance; trichome density associated with resistance	Cram (1978), Shanks and Doss (1986), Doss and Shanks (1988)
Needle nematode – *Longidorus elongatus*	Resistant genotypes were identified, segregation patterns suggest high heritability	Szczygiel (1981a)
Northern root-knot nematode – *Meloidogyne hapla*	Resistant genotypes were identified, segregation patterns suggest high heritability	Szczygiel (1981b), Szczygiel and Danek (1984), Edwards *et al.* (1985)
Obscure root weevil – *Sciopithes obscurus*	Tolerant genotypes were identified	Cram (1978)
Root aphid – *Aphis forbesi*	Resistant genotypes were identified	Darrow *et al.* (1933)
Root lesion nematode – *Pratylenchus penetrans*	Resistant genotypes were identified	Szczygiel (1981c), Potter and Dale (1994)

Table 4.1. *Continued*

Disease or pest	Observations	Representative studies
Strawberry root weevil – *Otiorhynchus* ssp.	Tolerant genotypes were identified	Cram (1978)
Woods weevil – *Nemocestes incomptus*	Tolerant genotypes were identified	Cram (1978)

Both leaf volatiles and essential oil content have been examined as possible inhibitors to two-spotted spider mite attack (Hamilton-Kemp *et al.*, 1980; Khanizadeh and Bélanger, 1997). Moderate to high levels of heritability have been found to grey mould (*Botrytis cinerea*, Pers. Ex. Fr.), leaf spot, leaf scorch, powdery mildew, ramularia leaf spot, red stele, viruses, verticillium wilt, strawberry aphid, black vine weevil, and two-spotted mite. Black leaf spot resistance has been reported to be at a single locus (Yamamoto *et al.*, 1985). It has been suggested that there are five virulence genes and five resistance genes to red stele (Van de Weg *et al.*, 1993, 1997; Van de Weg 1997).

Recently, considerable attention has been placed on resistance to soil pathogens, because of the impending ban on methylbromide. Screens of the California breeding population on fumigated and non-fumigated soil have uncovered little resistance to the typical soil pathogens found in strawberry fields as measured by either runner production (Larson and Shaw, 1995a) or productivity (Larson and Shaw, 1995b). Without fumigation cultivars perform more than 50% worse on average. Conventional wisdom suggests that eastern cultivars perform better on non-fumigated soils than California ones, but most of the comparative studies were done on soils where straw-berries had not been grown for long periods of time (Strang *et al.*, 1985; Gleason *et al.*, 1989). In comparisons on old strawberry sites, eastern geno-types have generally performed almost as poorly as California ones on non-fumigated soil (Keefer *et al.*, 1978; Shaw and Larson, 1996).

Breeding for resistant types has frequently been complicated by negative correlations between resistance and horticulturally important traits (Maas and Galletta, 1989; Hancock *et al.*, 1990). For example, Bringhurst *et al.* (1967) found verticillium wilt resistance was negatively correlated with yield. Breeding for disease resistance has been further complicated by the presence of eco- or biotypes of the pathogen. In the case of red stele root rot, there are over ten known races in the US, 12 in the UK, six in Canada and six in Japan (Pepin and Daubeny, 1964; Montgomerie, 1967; Morita, 1968; Maas *et al.*, 1988; Hancock *et al.*, 1997). Regional variation in cultivar susceptibility to pathogens has also been documented for alternaria leaf spot, ramularia leaf spot, leaf scorch, verticillium wilt and anthracnose (Hancock *et al.*, 1996b).

Range of adaptation

Strawberries are grown across a vast environmental zone and as a result, cultivars display variable adaptations to winter cold and spring frost (Bazzocchi, 1968; Daubeny *et al.*, 1970; Ourecky and Reich, 1976; Hummel and Moore, 1997). Cultivars grown in the more northern regions tend to be more winter hardy and this hardiness is highly heritable. A wide range in bloom tolerance to frost has also been described, although regional correlations were not always apparent (Oureky and Reich, 1976). Powers (1945) found that extreme winter hardiness was partially dominant over non-hardiness in breeding populations containing *F. virginiana* ssp. *glauca*, and that most of the important economic characters were independently inherited. Darrow and Scott (1947) found that *F. virginiana* ssp. *glauca* also transmits heritable tolerance to frost injury.

Although there are numerous published suggestions that cultivars vary in their resistance to heat and drought, there are few formal genetic studies. According to Darrow (1966), 'Blakemore' and 'Missionary' are among the most resistant cultivars to heat. Abdelrahman (1984) found reproductive effort in 'Pocahontas' to be less affected by high temperature (40°C/30°C day/night regimes) than 'Sparkle' and 'Raritan', although the relative rankings were different for vegetative responses.

Cultivars vary substantially in their chilling requirements, time of bloom and ripening dates. Few quantitative genetic studies have been performed on these characteristics, although Peterson (1953) did find that the dates of bloom and harvest were inherited quantitatively without heterosis. Cultivars that are adapted to warm southern areas, such as the southern USA, Mediterranean regions and Africa, appear to have the shortest rest periods and these plants are capable of growing and ripening fruit during the short days of summer (Piringer and Scott, 1964; Darrow, 1966; Lee *et al.*, 1968; Kronenberg and Wassenaar, 1972; Craig and Brown, 1977). There is also considerable variation in bloom and ripening dates within regions of adaptation, with time of bloom and ripening dates often being closely correlated (Wilson and Giamalva, 1954; Zych, 1966). Earliness can act as a partially dominant trait, as the F_1 hybrids of *F. virginiana* ssp. *glauca* × *F.* × *ananassa* were almost as early as the earliest parent (Powers, 1945). Scott *et al.* (1972) also noted that the earliness of *F. virginiana* behaved as a partial dominant.

Flowering and fruiting habit

Most commercial strawberries are now strict hermaphrodites, but sex appears to be a single gene trait subject to disomic inheritance in *F. vesca* ssp. *bracteata*, *F. chiloensis*, *F. virginiana* and *F.* × *ananassa* (Staudt, 1989; Ahmadi and Bringhurst, 1991; Irkaeva, 1993). Female (F) (pistillate) is dominant to hermaphrodite (H), which is dominate to male (M) (staminant). Females are heterogametic (F/H or F/M), wheras hermaphrodites can be homo- or heterogametic (H/H or H/M) and males are homogametic (M/M). A range in fertility can be found in hermaphrodites ranging from self-infertility to complete fruit set (Stahler *et al.*, 1990, 1995; Luby and Stahler, 1993). In *F. orientalis* and *F. moschata*, Staudt (1967a, b) found tetrasomic inheritance for sex and he described the alleles for sex as male suppressor Su^M (F) dominant to male inducer Su^+ (H) and to the female suppressor Su^F (M). Su^F was dominant to Su^+. Staudt (1997) also provides evidence that genes for male organs in *F. chiloensis* may be inactivated by the cytoplasm of *F. virginiana*.

There appear to be two types of 'everbearing' octoploid plants based on their photoperiods: day-neutrals and long-day; however continuums in growth habit and flowering behaviour make rigid classifications difficult (Clark, 1938; Nicoll and Galletta, 1987). Genes for the everbearing characteristic came from a number of sources. 'Climax' was one of the first everbearers to appear in Europe, even though it was a cross between short-day types (Hedrick, 1925) and does not perform as an everbearer in warm climates (Downs and Piringer, 1955). It has also been suggested that the everbearing trait may have been transferred from perpetual-flowering types of *F. vesca* into early European octoploid populations (Richardson, 1914; Lesourd, 1965). The original source used in USA breeding programmes was 'Pan American' (Sjulin and Dale, 1987), which was a chance seedling or clonal mutation of 'Bismark' found in New York by S. Cooper in 1898 (Hedrich, 1925). Most recently, day-neutrality was derived from native clones of *F. virginiana* ssp. *glauca* (Powers, 1945; Bringhurst and Voth, 1984). Clones from the Rocky Mountains were first used by Powers (1954) in combination with cultivars containing 'Pan American' genes, and in the last two decades, Bringhurst and Voth (1978) released day-neutral types that were derived by backcrossing into short day *F. × ananassa* with a single clone of *F. virginiana* ssp. *glauca* from the Wasatch Mountains of Utah.

The inheritance behaviour of multiple cropping in strawberries has been the subject of numerous studies (Table 4.2), although it is not clear whether the same genes were investigated in each study. In a study of early sources of day-neutrality without *F. virginiana* ssp. *glauca* in their background, Darrow (1937) crossed non-everbearers with everbearers and obtained progeny ratios of nearly 3 : 1, leading him to believe that the trait in his population was regulated by a recessive gene. When Macoun (1924) crossed two everbearers, he got approximately a 9 : 7 ratio, indicating that everbearingness may be due to two dominant complementary genes. Richardson (1914) obtained about a 2 : 1 ratio (non-everbearer to everbearer ratio) in crosses of everbearers to non-everbearers, suggesting that everbearing was partially dominant with complex interactions being involved. Clark (1938) found different everbearing parents generated differing proportions of everbearing progeny that fitted no consistent ratio, and concluded that the everbearing tendency behaved as a

Table 4.2. Studies investigating the genetics of everbearingness.

Investigator	Source of everbearing trait	Conclusions
Richardson (1914)	European *F.* × *ananassa* mutants	Partially dominant trait with complex interactions
Macoun (1924)	N. American? *F.* × *ananassa* mutants	Two dominant complementary genes
Darrow (1937)	N. American? *F.* × *ananassa* mutants	Single recessive allele
Clark (1938)	N. American *F.* × *ananassa* mutants	Dominant character with factor interaction
Powers (1954)	N. American *F.* × *ananassa* mutants and *F. virginiana* ssp. *glauca*	Two complementary dominant genes and four recessives
Ourecky and Slate (1968)	N. American *F.* × *ananassa* mutants and *F. virginiana* ssp. *glauca*	Two complementary dominant genes segregating octosomally
Bringhurst and Voth (1978)	*F. virginiana* ssp. *glauca*	Single dominant gene
Simpson and Sharp (1988)	European? *F.* × *ananassa* mutants	Dominant trait
Ahmadi *et al.* (1991)	*F.* × *ananassa* mutants	Single dominant gene

'dominant character affected by some type of factor interaction'. Some of his everbearers, however, did not produce any everbearing progeny when selfed.

In a breeding population containing genes from both 'Pan American' and *F. virginiana* ssp. *glauca,* Powers (1954) recovered a broad range of progeny ratios in parents with similar background and suggested that they best fit a model of two or more complementary dominant genes of equal potency and at least four recessive genes governing expression of the character. He suggested that the effect of these genes was cumulative, but that the dominant genes did not contribute equally. Ourecky and Slate (1967) also decided that two complementary dominant genes governed inheritance, and that they segregated in an octoploid fashion. Everbearing × everbearing crosses yielded 61–79% everbearers, whereas everbearers × non-everbearers yielded percentages varying from 24% to 47%. Their evidence indicated that some short-day plants carried genes for everbearingness. In both these studies, the everbearer 'Geneva' produced more everbearing progeny than those with *F. virginiana* ssp. *glauca* in their background. Most recently, Bringhurst and associates (Bringhurst and Voth, 1978; Ahmadi *et al.*, 1991) found progeny ratios to most closely fit a single gene, dominant model in their populations where day-neutrality was derived from just *F. virginiana* ssp. *glauca.* They also found the gene was expressed in hybrids with other *Fragaria* and *Potentilla* species.

Although most studies have identified one or more dominant genes associated with the everbearing trait, various levels of interaction with other genes have been reported, and in fact, most breeders have observed complex inheritance patterns, even when they use the same sources of day-neutrality as Bringhurst and Voth. The differences in these studies may be due to several possibilities including: (i) there may be multiple sources of day-neutrality now segregating in octoploid breeding populations: (ii) the genetic background of the individual breeding populations may be variable enough to influence the expression of day-neutrality; and (iii) many everbearers may be misclassified as short-day plants. Additive genes for earliness can play a particularly confounding role in recognizing day-neutrals (Barritt *et al.*, 1982), although early short-day parents do not always yield higher proportions of day-neutral progeny (Simpson and Sharp, 1988). Many day-neutral individuals also do not even express the everbearing characteristic until the second year (Barritt *et al.*, 1982; Ourecky and Slate, 1967; Richardson, 1914). Perhaps it is best to consider the inheritance of everbearingness as a quantitative trait with a few major genes and many minor ones. Interestingly, no homozygous everbearing genotypes have been reported.

It is also possible that the same major genes may control both day-neutral and long-day flowering responses in octoploid strawberries, with modifiers playing the distinguishing role. The source clones of *F. virginiana* ssp. *glauca* are not day-neutral; they continue flowering through long days once they are triggered by short days. Bringhurst and Voth (1978) suggest that the mechanisms which normally delay flowering in *F. virginiana* are 'confounded' by certain backgrounds of *F.* × *ananassa*. The necessary background appears to be a lack of a rest requirement and the low chilling trait. Hot summer days also appear to inhibit flower bud formation in most day-neutral types (Durner *et al.*, 1984).

The genetics of multiple cropping in 'alpine' forms of European *F. vesca* is much simpler than that of *F.* × *ananassa*, due partly to their diploid instead of octoploid nature. The everbearers 'Baron Solemacher' and 'Bush White' contain a homozygous recessive gene for day-neutrality (Brown and Wareing, 1965). In addition, they carry two recessive alleles for non-running at an independent locus, one not associated with plant form, and the other associated with bushiness. A phosphoglucose isomerase isozyme loci has been found to be loosely linked to the locus governing the runnerless trait (Yu and Davis, 1995). Day neutrality has not been observed in North American populations of *F. vesca*, and when California clones were crossed with alpine forms, at least three genes were identified that controlled photoperiodism (Ahmadi *et al.*, 1991).

Most day-neutral types of both diploids and octoploids produce limited numbers of runners, although Simpson and Sharp (1988) found considerable variation for stolon production and yield in everbearing, octoploid types. General combining ability was the strongest component of fruit yield, but specific combining ability played a more important role in stolon production. They suggested that early fruiting and adequate stolon production could be combined in an everbearing type.

Fruit quality

Size of fruit is inherited quantitatively (Powers, 1945; Baker, 1952; Comstock *et al.*, 1958; Scott, 1959; Hansche *et al.*, 1968; Spangelo *et al.*, 1971b), with six to eight allelic pairs controlling fruit expansion (Sherman *et al.*, 1966). There is a decline in size of fruits from the primary to inferior positions, but the relative decline varies substantially among genotypes (Valleau, 1918; Moore *et al.*, 1970; Pelofske and Lawrence, 1984). Several studies have shown that a large part of the genetic variance for fruit size is epistatic, although there is still considerable additive variability, depending on parent populations (Scott, 1959; Sherman *et al.*, 1966; Hansche *et al.*, 1968; Scott *et al.*, 1972).

Considerable genetic variability has also been documented in the firmness, colour, composition and appearance of fruit. Flesh firmness and skin toughness are often correlated positively (Ourecky and Bourne, 1968) and are generally inherited quantitatively (Morrow and Darrow, 1941; Hansche *et al.*, 1968; Barritt, 1979; Shaw *et al.*, 1987), although Spangelo *et al.* (1971b) found little additive variation for fruit firmness in their population. Skin and flesh colour are largely under the control of additive variation (Murawski, 1968; MacLachlan, 1974; Lundergan and Moore, 1975), with a few genes having comparatively large effects (Shaw and Sacks, 1995). Correlations between internal and external colour are small suggesting separate sets of genes (Shaw, 1991). Soluble solids and acidity are controlled with varying levels of additive and dominance control (Duewer and Zych, 1967; Lal and Seth, 1979; Shaw *et al.*, 1987; Shaw, 1988). Shaw (1988) found little difference in the soluble solids and total sugars in his breeding population, although he did observe significant genotypic variation in sucrose, glucose, fructose and acidity levels. Wenzel (1980) found a negative association between soluble solids concentration and yield. Vitamin C content has been shown to be polygenic (Hansen and Waldo, 1944; Lundergan and Moore, 1975; Lal and Seth, 1979), with some parents displaying partial dominance for high levels (Anstey and Wilcox, 1950) and some progeny having higher levels than their parents (Darrow *et al.*, 1947). Several studies have described variation in the flavour of progeny families suggesting additive quantitative control (Darrow, 1966; Overcash *et al.*, 1943; Slate, 1943), although few formal genetic studies on this character have been conducted. Zubov and Stankevich (1982) found significant seedling variation in fruit consistency, anthocyanin content and vitamin C, but not flavour. General combining ability (GCA) was greater than specific combining ability (SCA) for all the other traits except vitamin C.

Yield

Yield is the product of a combination of characters, such as number and size of fruit, plant vigour, hardiness, and disease resistance of the plant. Crown number per row area is often the factor most strongly associated with yield (Rogers and Modlibowska, 1951; Craig and Aalders, 1966; Hancock *et al.*, 1983), although flower number and fruit size are also important components (Hondelmann, 1965). High crown numbers can be achieved through either high levels of stolon production or branch crown production.

Strong compensatory interactions have often been found between the various yield components (plant density, crowns per plant, trusses per crown, fruit per truss, etc.) (Bedard *et al.*, 1971; Lacy, 1973; Webb *et al.*, 1974; Mason and Rath, 1980; Swartz *et al.*, 1985), indicating that breeding for high fruit numbers or individual fruit size by themselves will not necessarily increase productivity. However, outlier types do exist with both large fruit and high fruit numbers (Hancock and Bringhurst, 1988).

Considerable levels of genetic variability have been described for most yield components, although the relative levels of additive and non-additive variation have varied greatly from study to study. In a few breeding populations, non-additive gene influences have appeared to be more important than additive ones. Spangelo *et al.* (1971b) and Watkins *et al.* (1970) studied the genetic variance components of 20 characters in 64 progenies from crosses involving 32 parents. They found that non-additive variance comprised about 50% of the total genetic variance, and that much of this was epistatic. Since epistatic variance was so important, Watkins *et al.* (1970) postulated that progress could best be obtained by a two-step breeding procedure, which would involve small-scale testing of progenies, followed by growing large progenies of the best ones. In a related study, Spangelo *et al.* (1971b) found that heritability estimates were high for a number of yield components such as average berry weight, berries per flower stem, yield per flower stem, and flower stem number. However, Watkins and Spangelo (1971) did not obtain increased yields when only additive variance was exploited. They suggested that, when several characters are involved in a breeding programme, crosses should be designed to exploit all the genetic variance, whether it be additive, dominant, or epistatic.

In most other studies, sufficient levels of additive variation were considered available for rapid improvement of yield. Morrow *et al.* (1958) studied eight characters involving 40 parents and discovered that high levels of genetic variation existed for most of the characters, including yield. Using the same source of data, Comstock *et al.* (1958) found that epistatic variance accounted for a sizeable part of the total genetic variance for five characters, but that significant levels of additive and dominance variance were also present. The five characters investigated were size of plant area, number of berries, total weight of berries, average weight per berry and weight of berries per unit plot size.

Hansche *et al.* (1968) determined the genetic and environmental variances, heritabilities, and genetic and phenotypic correlations associated with fruit size, firmness, yield and appearance in the University of California (UC-Davis) breeding programme. Extensive genetic variability was found associated with all the characters. Unlike the studies by Watkins *et al.* (1970) and Spangelo *et al.* (1971b), high levels of additive variance were associated with yield. Fruit firmness and size were also highly heritable, although fruit appearance showed little additive variation. A significant genetic correlation existed between fruit size and yield, indicating that plants with large berries have a genetic potential for high yield. When the UC-Davis breeding population was evaluated 20 years later, heritability estimates were not significantly different from the ancestral population (Shaw *et al.*, 1989). Environment had a large effect on estimates of genetic parameters, indicating a potential bias in the studies at other locations.

Other studies documenting high heritabilities for yield and its component parts include work by Aalders and Craig (1968), Bedard *et al.* (1971), Lal and Seth (1979, 1981, 1982) and Wenzel (1980). Aalders and Craig used seven inbred lines in diallele crosses to study inheritance of yield, appearance, quality, and mildew resistance. Significant general combining ability occurred for all four characters, and specific combining ability existed for yield, appearance, and mildew incidence. Bedard and coworkers studied the correlations among 28 fruit and plant characters. Total berry number and total marketable yields were positively correlated with average berry weight, berries per flower stem, yield per flower stem, leaf area and petiole diameter, but negatively correlated with stolon number and flower stem number. They concluded that no genetic barrier existed to prevent combining high yield with good berry quality. Wenzel (1980) examined phenotypic and genotypic correlations between yield, berry size, freezing ability and soluble solids. He found that the development of selection indices helped to identify elite progeny for all traits, but did not necessarily help to predict yield advances.

Lal and Seth (1979, 1981, 1982) did a series of studies that partitioned variability in a large number of yield associated traits into genotypic and environmental contributions. They discovered significant additive and non-additive effects for a wide array of traits including flower and fruit numbers, fruit yield and ripening dates. Among these traits, general combining ability was greater than specific combining ability, although a few traits such as fruit weight and runner number showed primarily non-additive regulation. They also observed cytoplasmic effects for many traits.

Numerous studies have been designed to test the effects of temporal, spatial and developmental variation on production traits, with the final intent to develop efficient selection strategies. Year to year, plot and site variability play key roles in phenotypic variation. Hortynski (1989) measured genotype–environment interactions for yield and fruit size using three F_1 populations. They found yield was more greatly affected by genotype × year interactions than genotype × block interactions. The opposite was the case for fruit size. Gooding *et al.* (1975) also found that site had a major effect on inflorescences per crown, but not fruit size. Shaw and coworkers found that within a single year, the distributions of genetic and environmental variance components for a single trait vary continuously, with heritabilities for yield and fruit size being highest in the middle of the season. They also found that seedling location has a large effect on the expression of genetic variation (Shaw, 1989; Shaw *et al.*, 1989). Shaw (1991) observed that 'cultural treatments that induce variable competition generate large G×E interaction for yield components, but the consequences for yield are smaller due to compensation', and 'Yearly variation induced large average performance differences, but relatively small interactions'. In annual systems, nursery treatments induced large interactions for production traits, especially those that condition variable levels of plant development and chilling. His studies indicate that choosing crosses among parents chosen for breeding value may be more effective than simple clonal performance in generating superior seedling populations. In fact, the performance of seedlings may be very different when propagated as runner plants or when grown in different environments.

Adaptability to mechanical harvesting

A recurring objective in strawberry breeding is to produce types adapted to mechanical harvesting. Paramount are concentrated ripening for once over harvest, long pedicles and either easy calyx removal or long necks for machine decapping (Sistrunk and Moore, 1980; Gooding *et al.*, 1983; Dale and Hergert, 1991). There is considerable variation for concentrated ripening (Denison and Buchele, 1967; Moore *et al.*, 1970; Barritt, 1974), although concentrated ripeners are often lower yielding than longer-season types (Moore *et al.*, 1975). Ease of calyx removal shows considerable additive genetic action (Barritt, 1976), and in some parents, low capping force is dominant (Brown and Moore, 1975). Considerable variation in pedicle and fruit neck length have also been reported (MacIntyre and Gooding, 1978; Dale *et al.*, 1987). Unfortunately, genotypes have not been developed to date, that produce consistently profitable yields when machine harvested (see Chapter 6).

Variegation

Variegation (June yellows or transient yellows) is an abnormal mottling appearance that has appeared in a number of breeding populations and cutivars (Darrow, 1955; Hughes, 1989). Symptoms vary from light green–yellow mottling to intense green and yellow mottling. Variegation can occur at all ages of a clone, from young seedling to long-propagated cultivar (Rose, 1992). Notable examples of the appearance of June yellows in cultivars long after release are 'Blakemore', 'Howard 17', 'Tufts', 'Goldsmith' (Z5A), 'Glooscap' and 'Blomidon' (Jamieson and Sanford, 1996). Inheritance of these variegations do not fit a Mendelian pattern (Williams, 1955; Wills 1962;

Misic, 1995) and cytoplasmic control is, therefore, the likely cause of the disorder (Rose, 1992). All attempts to transmit a causal virus have failed.

BIOTECHNOLOGICAL APPROACHES TO GENETIC IMPROVEMENT

Strawberries were one of the first crops to be routinely proliferated through micropropagation, but molecular work on them is only beginning to emerge in quantity. Regeneration systems have been developed by several laboratories using leaf discs (Liu and Sanford, 1988; Nehra *et al.*, 1989), callus (Nishi and Oosawa, 1973; Wang *et al.*, 1984 Jones *et al.*, 1988; Miller and Chandler, 1990), protoplasts (Nyman and Wallin, 1992a, b), and petiole cuttings (Graham *et al.*, 1995). Protoplasts of *F.* × *ananassa* and *F. vesca* have been successfully fused in culture (Wallin, 1997), and plants have been regenerated from cotyledons of *F. vesca* × *Potentilla fruticosa* hybrids (Silva and Jones, 1996). Haploids have also been produced from several octoploid cultivars (Owen and Miller, 1996). In a few cases, somaclonal variation has been uncovered that appeared to be horticulturally useful (Yurgalevitch *et al.*, 1985; Popescu *et al.*, 1997).

Strawberry leaf pieces and petiole sections have been genetically transformed with disarmed strains of *Agrobacterium tumifaciens* (James *et al.*, 1990; Nehra, *et al.*, 1990; Graham *et al.*, 1995), and electroporation has been used to transfer the GUS gene to strawberry protoplasts (Nyman and Wallin, 1992b). Several genes have been incorporated using *Agrobacterium*, including the *S*-adenosylmethionine hydrolase (SAMase) gene which controls ethylene biosynthesis (Mathews *et al.*, 1995), cowpea trypsin inhibitor which has been implicated in insect resistance (Graham *et al.*, 1995, 1997), and phosphinothricin-specific *N*-acetyl transferases (PAT enzyme) which confers herbicide resistance (du Pressis *et al.*, 1995). A number of efforts are being made to develop coat protein-mediated resistance, including SMYEPV and several nematode-borne viruses in Europe (Martin, 1995). Another gene being targeted for insect resistance is polygalacturonase-inhibiting protein (PGIP) from immature raspberry fruits (Williamson *et al.*, 1993; Ellis, 1995).

There are only a few reports of strawberry gene characterization, although Manning (1997, 1998) has developed a library of fruit-ripening genes for this purpose. Wolyn and Jelenkovic (1990) have cloned and sequenced the 18S rRNA and alcohol dehydrogenase genes of strawberry. A number of groups have used isozymes (Bringhurst *et al.*, 1981; Bell and Simpson, 1994), the polymerase chain (PCR) reaction and random primers to characterize elite selections and cultivars (Hancock *et al.*, 1994; Parent and Pagé, 1995; Graham *et al.*, 1996; Degani *et al.*, 1998). Random amplified polymorphic DNA (RAPD) markers have been found that are tightly linked to red stele resistance genes (Haymes *et al.*, 1997), and Battey *et al.* (1998) are searching for linkages to the day-neutrality gene in *F. vesca* using inter-simple sequence repeat-polymerase chain reaction (ISSR–PCR). Davis and his group have used a number of RAPD and isozyme markers to develop a genetic map of *F. vesca* (Davis and Yu, 1997), find genetic linkages with fruit colour (Williamson *et al.*, 1995) and runnering (Yu and Davis, 1995).

UTILIZATION OF NATIVE GERMPLASM

Interest has been growing in using native germplasm in breeding programmes (Hancock and Luby, 1993; Hancock *et al.*, 1993). This interest was undoubtedly spurred by the recent introduction of the day-neutral trait by Bringhurst and Voth, but it has also been encouraged by breeders realizing that the genetic base of modern cultivars is surprisingly narrow. The majority of the genetic makeup of the cultivars released since 1990 comes from only seven nuclear and 10 cytoplasmic sources (Sjulin and Dale, 1987; Dale and Sjulin, 1990). Such a narrow genetic base is of concern due to the possibility of detrimental inbreeding effects, lost opportunities and a lack of diversity to face new environmental challenges.

Numerous valuable characteristics exist in the lower ploidy species that could be of value in the cultivated species (Darrow, 1966; Hancock, 1990). *Fragaria iinumae*, *F. vesca* and *F. nipponica* are located on cold, alpine meadows. *F. vesca* has high tolerance to heat and drought, and high aroma along with resistance to verticillium wilt (Arulsekar, 1979), powdery mildew (Harland and King, 1957) and crown rot (*Phytophthora cactorum*) (Gooding *et al.*, 1981). *Fragaria moschata* is found under heavy shade and is immune to powdery mildew (Maas, 1998). *Fragaria viridis* tolerates alkaline soils. In a comprehensive study of diploid species in Ontario, Bors and Sullivan (1998) found *F. nilgerrensis* to have immunity to aphids and leaf diseases. *F. iinumae* produced unusual tap roots from runners. *F. moschata* survived a particularly cold winter in water-logged soil and displayed excellent leaf disease resistance. *Fragaria pentaphylla* was extremely vigorous, with unusually bright red, firm fruit and leaf disease immunity.

The incorporation of traits from a number of lower ploid species has been accomplished through pollinations with native unreduced gametes or by artificially doubling chromosome numbers (Hancock *et al.*, 1996b). The utility of this approach has been shown for a wide range of species in *Fragaria* and in the related genus *Potentilla* (Ellis, 1962; MacFarlane Smith and Jones, 1985; Ahmadi and Bringhurst, 1992; Noguchi *et al.*, 1997; Trajkovski, 1997). Particular success in incorporating lower ploidies into the background of *F. × ananassa* has come through combining lower ploidy species and then doubling to the octoploid level (Sangiacomo and Sullivan, 1994; Bors and Sullivan, 1998).

Native clones of *F. chiloensis* and *F. virginiana* also offer a rich genetic storehouse (Tables 4.3 and 4.4) and may be more efficiently used in improving *F. ×*

ananassa than the lower ploidies, as they cross readily with the cultivated types and still offer much genetic diversity. Some of the wild clones have particularly interesting flavours and aromas that have not yet been characterized, and they possess resistance to extreme environments, as well as a number of disease and pest problems. In addition, variability exists in several yield-related physiological traits including: (i) heat and cold tolerance, (ii) rates and patterns of CO_2 fixation, (iii) the levels of dry matter allocated to reproduction, (iv) the number of flowering cycles; and (v) the length of the floral induction period. In many cases, important components of yield can be combined with known disease resistances.

Most reports of pest resistance are limited to one disease or insect, but there are some genotypes that have been identified with multiple resistances. Among the most impressive clones of *F. chiloensis* are BSP 14, CA 11, DL 40, LCM 10, RCP 37 and TR 4 which carry resistance to aphids, two-spotted spider mites, red stele, leaf spot and powdery mildew. RCP 37 and CA 11 are also resistant to root lesion nematodes, have very high photosynthetic rates, and originated on dry, salty dunes (Hancock *et al.*, 1989a, 1996a).

Although backcrossing strategies have usually been employed in utilizing native strawberry germplasm, an effective strategy might also be to reconstitute *F. × ananassa* by crossing elite, complementary clones of *F. virginiana* with *F. chiloensis*, particularly when the native germplasm carries more than one useful trait (Hancock *et al.*, 1993). This approach would more effectively utilize the great amount of diversity available, and could result in exciting, new combinations of genes that would far outweigh the added time investment. The initial hybrids with *F. × ananassa* often have low yield and small fruit size, but exceptions do occur (Bringhurst and Voth, 1960), and good horticultural types can be re-covered within three generations of backcrossing to the cultivated parent (Bringhurst and Voth, 1984).

CURRENT BREEDING EFFORTS

There are numerous public and private breeding programmes across the world. The largest European efforts are found in France, Italy, the Netherlands, Norway, Spain and the UK. Active programmes are also located in Denmark, Finland, Germany, Hungary, Poland, Sweden, Russia and Romania (Rosati, 1991). In France, the Centre Interrégional de Recherche et d'Expérimentation de la Fraise (CIREF) directed by P. Roudeillac and M. Markocic (1997) is active in developing short-day and day-neutral cultivars with superior fruit appearance and flavour as well as soil disease tolerance. In Italy, there is a national programme 'Frutticoltura' funded by the Minister of Agriculture, with 17 institutions involved (Faedi *et al.*, 1997). Concentration is being placed on developing new dessert varieties with adaptations to the south, Po

Table 4.3. Horticulturally useful traits screened in wild *F. chiloensis*.

Trait	Collection site[a]	Reference
Morphological		
Flower/fruit numbers	CA, PNW, SA	Hancock and Bringhurst (1979b, 1988), Luby *et al.* (1991), Lavín (1997), M. Luffman, Ontario, 1988, personal communication
Fruit colour	PNW, SA	Cameron *et al.* (1991, 1993), Lavín (1997), Luffman and Macdonald (1993)
Fruit size	CA, PNW, SA	Bringhurst and Voth (1960), Hancock and Bringhurst (1979b, 1988), Cameron *et al.* (1991, 1993), Luby *et al.* (1991), Luffman and Macdonald (1993), Lavín (1997)
Physiological		
Bloom date	CA, SA	Scott *et al.* (1972), Hancock and Bringhurst (1979b), Luby *et al.* (1991), Lavín (1997)
Fruiting date	CA, PNW, SA	Hancock and Bringhurst (1979b), Luffman and Macdonald (1993), Lavín (1997)
Duration of fruiting	CA, SA	Hancock and Bringhurst (1979b), Lavín (1997)
Soluble solids	SA, PNW	Cameron *et al.* (1991, 1993), Lavín (1997), M. Luffman, Ontario, 1988, personal communication
Nutrient uptake	CA, SA	Moon *et al.* (1990), Lavín (1997)
Photosynthetic rate	CA, PNW, SA	Hancock *et al.* (1989a), Cameron and Hartley (1990), Lavín (1997)
Drought tolerance	CA	Zhang and Archbold (1993a, b)
Salt tolerance	CA	Hancock and Bringhurst (1979b), Wright and Hughes (1993)
Winter cold tolerance	CA, PNW, SA	Luby *et al.* (1991)
Disease resistance		
Botrytis fruit rot	PNW	Luffman and Macdonald (1993)
Leaf scorch	PNW	Luffman and Macdonald (1993)
Powdery mildew	PNW	Luffman and Macdonald (1993)
Red stele	CA, PNW	Pepin and Daubeny (1966), Luby *et al.* (1991)
Verticillium wilt	CA, PNW, SA	van Adrichem and Orchard (1958), Bringhurst *et al.* (1966), Pepin and Daubeny (1966), Arulsekar (1979), Maas *et al.* (1989), Cameron *et al.* (1993)
Virus	CA, PNW, SA	Miller and Waldo (1959), Luby *et al.* (1991)

Table 4.3. *Continued*

Trait	Collection site[a]	Reference
Pest resistance		
Aphid	CA, PNW	Shanks and Barrit (1974), Crock *et al.* (1982), Luby *et al.* (1991)
Black vine weevil	CA, PNW	Shanks *et al.* (1984), Shanks and Doss (1986), Doss *et al.* (1987), Luby *et al.* (1991)
Spider mite	CA, PNW, SA	Shanks and Barrit (1984), Luby *et al.* (1991), Shanks and Moore (1995)
Root weevil	CA, PNW, SA	Luby *et al.* (1991)
Root lesion nematode	CA, PNW	Potter and Dale (1994)

[a]Abbreviations: CA, California; PNW, Pacific Northwest; SA, South America.

Table 4.4. Horticulturally useful traits screened in wild *F. virginiana*.

Traits	Collection site[a]	Reference
Morphological		
Flower numbers	CA, MW, NE, OT, PNW, RM, SE, UNW	Hancock and Bringhurst (1979b), Dale *et al.* (1992), Luby and Stahler (1993), Dale (1994); Sakin *et al.* (1997), J. Ballington, North Carolina, 1997, personal communication
Fruit size	MW, NE, OT, PNW, RM, SE, UNW	Stahler (1990), Dale *et al.* (1993), Luby and Stahler (1993), Dale (1994), Sakin *et al.* (1997), J. Ballington, North Carolina, 1997, personal communication
Fruit colour	MW, NE, OT, PNW, RM, SE, UNW	Stahler (1990), Luby *et al.* (1991), Dale *et al.* (1992), Luby and Stahler (1993), Dale (1994), J. Ballington, North Carolina, 1997, personal communication
Firmness	OT, SE	Scott (1959), Dale *et al.* (1992), Luby and Stahler (1993), Dale (1994), J. Ballington, North Carolina, 1997, personal communication
Physiological		
Bloom date	MW, NE, OT, PNW, SE, RM, UNW	Darrow and Scott (1947), Scott *et al.* (1972), Stahler (1990), Luby *et al.* (1991), Dale *et al.* (1992), Luby and Stahler (1993), Dale (1994), J. Ballington, North Carolina, 1997, personal communication
Everbearing	MW, NE, OT, PNW, RM, UNW	Powers (1954), Dale *et al.* (1992), Luby and Stahler (1993), Dale (1994), Sakin *et al.* (1997)

Table 4.4. *Continued*

Trait	Collection site[a]	Reference
Photosynthetic rate	RM	Hancock *et al.* (1996a)
Winter cold tolerance	MW, RM, NE, PNW; RM; UNW	Hildreth and Powers (1941), Powers (1945), Scott *et al.* (1972), Luby *et al.* (1991), Luby and Stahler (1993)
Disease resistance		
Angular leaf spot	MW, SE	Maas *et al.* (1999)
Anthracnose	SE	J. Ballington, North Carolina, 1997, personal communication
Crown rot	MW, RM, NE, PNW; SE, RM; UNW	Harrison *et al.* (1996), J. Ballington, North Carolina, 1997, personal communication
Leaf scorch	MW, NE, OT, PNW, RM, UNW	Luby and Stahler (1993), Dale (1994)
Powdery mildew	MW, NE, OT, PNW, RM, UNW	Luby and Stahler (1993), Harrison *et al.* (1996)
Verticillium wilt	MW, NE, RM	van Adrichem and Orchard (1958), Varney *et al.* (1959); Gooding (1972)
Pest resistance		
Spider mite	NE, MW, PNW, RM, SE	Shanks and Moore (1995)
Root lesion nematode	OT	Potter and Dale (1994)

[a]Abbreviations: CA, California; MW, midwestern US; NE, northeastern US; OT, Ontario; PNW, Pacific northwest (Oregon, Washington and British Columbia); RM, Rocky Mountains; SE, southeastern US; and UNW, upper northwest (northwestern Canada and Alaska).

Valley and northern mountain regions including disease resistance and tolerance to alkaline soils. A private company, Consorzio Italiano Vivaisti (C.I.V.) directed by M. Leis and D. Musacchi is also active in producing dessert varieties for both north and south Italy. The breeding effort of I.V.T. (CPRO-DLO) of the Netherlands, currently directed by B. Meelenbroek, is concerned with developing types with broad adaptations, high yields and large fruit size. The most important European cultivar, 'Elsanta' (1981) came from this programme. In Norway, A. Nes at the Norwegian Crop Research Institute is breeding for large, general-purpose berries that are early and of high quality (Nes, 1997). In Spain, a public breeding programme, coordinated by R. Bartual of the Instituto Valenciano de Investigaciones (IVIA) is concentrating on producing cultivars for the two major production areas, Valencia and Huelva. Their goals are to breed high yielding, early cultivars for the fresh market with large, high quality fruits (Bartual *et al.*, 1997). The private firm, Plantas de Navarra S.A. (PLANASA) directed by D. Sanchez Eguialde (1997) is also very active in searching for highly productive, high quality types for

Spain. In the UK, D. Simpson of Horticulture Research International, is concentrating on season extension using both short-day and day-neutral cultivars (Simpson *et al.*, 1997a). R.J. McNicol at the Scottish Crop Research Institute in Invergowrie is developing disease-resistant cultivars which are consistent cropping and need minimal agrochemical imputs (McNichol *et al.*, 1997). This institute has been actively breeding cultivars for over 50 years with Reid's 'Auchincruive Climax' (1947), 'Redgauntlet' (1957) and 'Talisman' (1955) being some of its most popular releases. The famous breeding programme at the University of Cambridge, where D. Boyles produced the famous Cambridge series of cultivars, has been eliminated.

Numerous breeding programmes are also found in North America, but their numbers have dwindled over the last several decades. There are now only about eight experiment stations doing breeding work in the US; this is less than half the number reported by Darrow in his 1966 monograph on the strawberry. State-supported programmes are currently located in Washington (P. Moore), Michigan (J. Hancock), Minnesota (J. Luby), New Jersey (J. Fiola and G. Jelenkovic), New York (C. Weber), North Carolina (J. Ballington), Wisconsin (B. Smith), and Florida (C. Chandler). Some of the most important cultivars that have come out of these programmes over the years are 'Honeoye' and 'Jewel' (D. Ourecky and J. Sanford, New York), 'Raritan' (C. Chandler, New Jersey) and 'Sweet Charlie' (L. Hough, Florida).

The current goals of the state-supported programmes are: (i) Florida – high fruit quality including resistance to water damage, high late November to mid-March yields, anthracnose resistance; (ii) Michigan – day-neutral types with higher heat tolerance and resistance to soil pathogens, germplasm development using wild octoploids; (iii) Minnesota – winter hardiness, high quality, disease resistance, germplasm development using wild octoploids; (iv) New Jersey – early cultivars with excellent fruit flavour and size that are adapted to both matted row and hill culture; (v) New York – fruit quality including size, symmetry; high, steady yields; black root rot resistance; (vi) North Carolina – superior genotypes that are resistant to anthracnose and are adapted to either annual hill or matted rows; (vii) Washington – June bearing types with firm, easily harvested fruit for both processing and fresh outlets, increased disease and insect resistance (particularly fruit rots and aphid transmitted viruses); and (viii) Wisconsin – germplasm development, high quality, high yielding, disease resistant types.

There are several large breeding programmes in Canada that are federally funded (Daubeny, 1990). The long-term programme of H. Daubeny in British Columbia is now being directed by C. Kempler. The emphasis there has traditionally been high-quality fresh and processed market cultivars with disease and pest resistance. In Nova Scotia, A. Jamison is making wide use of European cultivars to produce early season, red stele-resistant types. He is building on the previous decades of work of D.L. Craig and L.E. Aadlers. Some of the most significant cultivars to come out of their research are 'Bounty' (1972),

'Glooscap' (1983) and 'Kent' (1981). The Ontario programme led by A. Dale at Simcoe is concentrating on large-fruited, firm types and he is actively exploiting native germplasm. S. Khanizadeh and D. Buszard at McGill University and the Horticultural Research and Development Centre in Quebec are searching for large-fruited, paleskin-coloured and firm types with resistance to red stele.

The largest US breeding programmes reside at the University of California at Davis, and the USDA centred at Beltsville, Maryland and Corvallis, Oregon. Now under the direction of D. Shaw, the Cal-Davis programme, run for decades by R.S. Bringhurst and V. Voth, is known for its broadly adapted, large-fruited and high-yielding cultivars. An amazing succession of internationally important cultivars have been produced by the California programme including 'Tioga' (1964), 'Tufts' (1972), 'Aiko' (1975), 'Douglas' (1979), 'Pajaro' (1979), 'Chandler' (1983), 'Selva' (1983), 'Oso Grande' (1987), 'Camarosa' (1992), 'Seascape' (1991) and 'Gaviota' (1996). 'Camarosa', 'Chandler' and 'Selva' now dominate world acreage.

The USDA programme is probably the longest continually maintained programme in the world. Cultivars derived from this programme have been widely planted in the eastern US since 1920 through the efforts of G.M. Darrow, D.H. Scott, A.D. Draper, G.J. Galletta and, most recently, S. Hokanson. Some of the most important cultivars released from this programme are 'Blakemore' (1928), 'Pocahontas' (1950), 'Albritton' (1951), 'Surecrop' (1956), 'Sunrise' (1956), 'Redchief' (1968), 'Earliglow' (1975), 'Allstar' (1981), and 'Tribute' and 'Tristar' (1981). These cultivars are particularly noted for their resistance to soil pathogens. The USDA breeding work at Corvallis has been conducted through the years by G.F. Waldo, F.J. Lawrence and now C. Finn. Some of the more important cultivars which have emerged from this programme were 'Siletz' (1955) and 'Hood' (1965). Current emphasis is being placed on developing high quality types for the processed market.

Tom Sjulin (Research Department Manager) and Bruce Mowrey (Head Plant Breeder) of the Driscoll Strawberry Associates in Watsonville, California, direct the most vigorous private, breeding effort in the US. Their primary goals focus on consumer attributes of flavour, appearance and shelf life, coupled with the production attributes of fruit size, timing of harvest and harvestability. Other private efforts are conducted by Plant Sciences Inc. (California), New West Fruit Corporation (California) and Well-Pict Inc. (California).

Significant breeding work is also being conducted in Japan, Israel and Turkey. The largest number of public and private breeding programmes found anywhere are in Japan, where dozens of cultivars have been released in the last decade. Probably the largest programme is conducted by the National Research Institute of Vegetables, Ornamental Plants and Tea with two branches at Kurume and Morioka, but numerous other Prefecture Experiment Stations and private companies are active. Common goals are to produce large, dessert quality berries that are adaptable to forcing culture. In Israel, S. Izar and E. Izsak at the Volcani Centre, Bet Dagan, have made significant strides in developing cultivars that are very early, and are adapted to low elevations, mild winters and the absence of a chilling requirement. Breeding in Turkey is relatively new, but a major effort has been established at Çukurova University utilizing a combination of Turkish, European, and Californian varieties as breeding material (Kaşka, 1997).

CONCLUSIONS

Dramatic improvements have been made in the yield and quality of the strawberry over the last two centuries. Since the initial work of Thomas A. Knight in 1817, hundreds of improved cultivars have been developed across the world. Some of the most widely grown cultivars of the second half of the 20th century have come out of California ('Tioga', 'Camarosa', 'Chandler', 'Oso Grande' and 'Selva'), Nova Scotia ('Glooscap' and 'Kent'), New York ('Honeoye' and 'Jewel'), Maryland ('Allstar', 'Earliglow', 'Redchief', 'Tribute'), The Netherlands ('Elsanta' and 'Gorella') and the UK ('Cambridge Favourite', 'Redgauntlet').

These successes have come not only through the ingenuity of individual breeders, but also through the creative cultural research of plant physiologists. A large part of the dominance of the California strawberry industry is due to the development of a total cultural system that maximizes the potential of each individual cultivar. This system has been successfully modified to fit all warm winter climates.

Strawberry breeders have had the advantage of working with a heterozygous, asexually propagated crop that contains a large storehouse of genetic variability. The dessert strawberry *F. × ananassa*, is itself a hybrid of two species, *F. chiloensis* and *F. virginiana*, which are highly variable themselves and contain a number of complementary traits. Considerable effort has been undertaken to discover horticulturally useful traits in these species, and to determine their genetics. Some of the most dramatic breeding successes have come in the 20th century through the incorporation of genes for an extended fruiting season from 'Nich Ohmer' into California cultivars in the 1930s, and the more recent transfer of genes for day-neutrality from a native plant of *F. virginiana* ssp. *glauca* into *F. × ananassa*.

It seems likely that high levels of breeding success will continue, as long as adequate funding is available for research. Huge levels of untapped diversity remain in both breeding and natural populations, and a wide range of genes have been cloned from other species that will have a positive impact on strawberry productivity and quality, as they are added to the strawberry genome through biotechnological approaches. The danger is that funding for both conventional and non-conventional improvements will be increasingly targeted to the largest production areas, where the highest return on investment can be made. It is hoped that public funding will remain available in the other vast areas, where millions of other people still grow and enjoy strawberries.

5

STRUCTURAL AND DEVELOPMENTAL PHYSIOLOGY

INTRODUCTION

Commercial strawberries are successfully grown in a broad range of climates including temperate, grassland, Mediterranean, taiga and subtropical. However, most of the current production is limited to the temperate and Mediterranean climates located between latitudes 28° and 60° (Hancock *et al.*, 1996b). In these areas, average mid-summer temperatures in July are in the range of 15–30°C, with average summer highs of 20–40°C. Average mid-winter temperatures in January range from 15 to –20°C, with mean lows of 10 to –40°C (Galletta and Bringhurst, 1990; Hancock and Luby, 1995).

Strawberry development is regulated by a complex set of environmental and physiological cues. The appearance and growth of leaves, crowns, roots, runners and inflorescences is markedly influenced by genotype, and several environmental factors including temperature, light intensity and light quality/photoperiod (Guttridge, 1985; Durner and Poling, 1988; Larson, 1994; LeMière *et al.*, 1996).

ANATOMY AND MORPHOLOGY

The strawberry is a herbaceous perennial that has a central stem or crown from which leaves, roots, stolons (runners) and inflorescences emerge (Fig. 5.1). The crown consists of a central core surrounded by a vascular ring. The core is composed primarily of pith, with a thin cambial layer surrounding it. At the top of each leaf along the crown is an axillary bud, which can produce runners, branch crowns or remain dormant, depending on environmental conditions (Fig. 5.2).

The leaves are arranged spirally; each sixth leaf is above the first (Fig. 5.3). The leaves are generally pinnate and trifoliate. They have the epidermis, palisade and mesophyll layers typical of dicots. Stomata are only on their

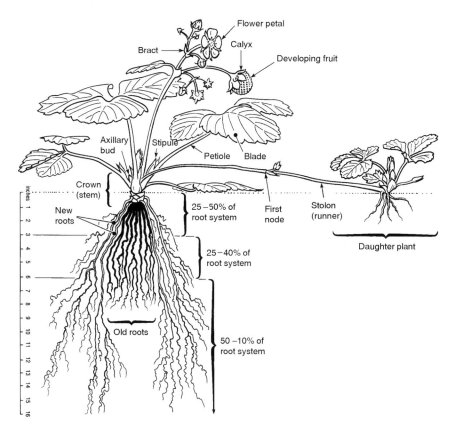

Fig. 5.1. A developing strawberry plant (reprinted by permission from Strand, 1994).

undersides (Darrow, 1966). Leaves of most species live only a few months and die after exposure to hard frosts in the autumn, although some leaves of *F. chiloensis* remain green throughout the winter if temperatures do not drop substantially below zero. Old leaves are replaced in the spring with new leaves that have overwintered as primordia between the protective layers of the stipules. In a vegetative bud, there are usually five to ten leaf primordia.

Roots emerge from the base of the crown where it comes in contact with the soil. The root anatomy is typical of dicots. Adventitious roots arise from the crown in the pericycle, and push out through the cortex (Fig. 5.1). The roots begin branching at 2–5 cm and if adequate water is available, they will keep branching into a fibrous mass. Generally, there are 20–30 primary roots, and there are hundreds of secondary, tertiary and higher orders roots. Up to 50–90% of the roots are concentrated in the upper 10–15 cm of the soil (Dana, 1980). Lateral roots live for 1 or 2 years, whereas primary roots can live for 2–3 years depending on species and environmental conditions. Roots

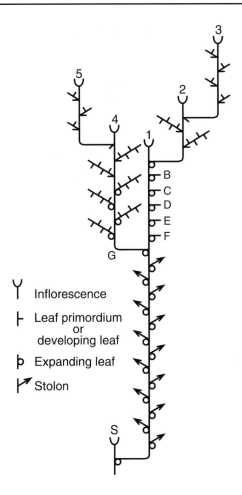

Fig. 5.2. A schematic drawing of a strawberry crown (redrawn from Guttridge, 1955).

are colonized by vesicular-arbuscular mycorrhizal species (Darrow, 1966; Khanizadeh *et al.*, 1995).

    Stolons of most species consist of two nodes (Fig. 5.1). A 'daughter plant' is produced at the second node, whereas the first node remains dormant or develops another stolon. Each daughter plant has the capacity to produce its own stolons. A vigorous plant of *F.* × *ananassa* usually produces 10–15 stolons a year, whereas a clone of *F. virginiana* can produce two or three times that number. Stolons of *F.* × *ananassa* and *F. virginiana* survive for as long as 1 year, whereas those of *F. chiloensis* can last for several years. The mother plant can transfer water, nutrients and assimilates to the daughter plant for several weeks to years depending on the genotype and species (Darrow, 1966; Alpert, 1991, 1996). Rooted daughter plants can usually survive independently after

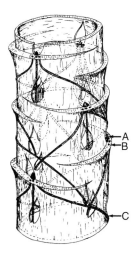

Fig. 5.3. Spiral arrangement of leaves along the crown (redrawn from Darrow, 1966). A = bud; B = leaf base; C = vascular strand.

2 or 3 weeks of attachment.

The strawberry inflorescence is a modified stem or cyme, terminated with a primary blossom (Fig. 5.1). Following the primary blossom, there are typically two secondaries, four tertiaries and eight quaternaries (Fig. 5.4). A typical blossom has ten sepals, five petals and 20–30 stamens (Fig. 5.5). Pistils number from 60 to 600. The greatest number of pistils are found in the primary blossoms, decreasing successively in number from primaries to quaternaries.

Strawberries are insect pollinated, with bees being used most commonly as pollinators.

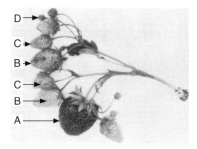

Fig. 5.4. Arrangement of the strawberry fruiting cluster. (A) primary fruit; (B) secondary fruit, (C) tertiary fruit; (D) quaternary fruit (redrawn from Darrow, 1966).

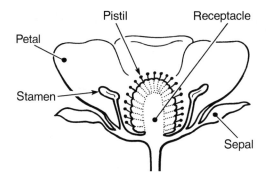

Fig. 5.5. Typical strawberry blossom (reprinted by permission from Strand, 1994).

Pollen matures before the anthers open, but it is not dehisced until the flowers open (Darrow, 1966; Hancock *et al.*, 1996b). It remains viable for 2–3 days. Stigmata remain receptive to pollen for 8–10 days. Fertilization occurs 24–48 h after pollination.

The fruit of the strawberry is an 'aggregate', composed of numerous ovaries, each with a single ovule (Fig. 5.6). The resulting seeds are called 'achenes' and are the true fruit of the strawberry. The embryo consists of two large, semielliptical cotyledons, which contain protein and fat, but no starch (Darrow, 1966). The receptacle is composed of an epidermal layer, a cortex and a pith. The latter two layers are separated by vascular bundles that supply nutrients to the developing embryos.

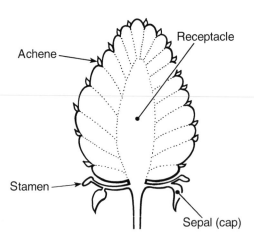

Fig. 5.6. Typical aggregate fruit of strawberry (reprinted by permission from Strand, 1994).

PHOTOPERIODIC AND TEMPERATURE REGULATION OF DEVELOPMENT

There are two primary types of strawberries now grown commercially, day-neutral and short-day plants. Long-day ('everbearing') plants are also avail-able, but they are rarely grown outside of home gardens. The short-day types are actually facultative short-day plants and initiate flower buds either under short-day conditions (less than 14 h of day length; Fig. 5.7) or when temperatures are less than 15°C (Fig. 5.8; Darrow, 1936; Guttridge, 1985; Larson, 1994). Above 15°C, the critical photoperiod for floral induction is 8–12 h, depending on the cultivar. Long-day plants typically initiate their flower buds when day lengths are greater than 12 h, and temperatures are moderate.

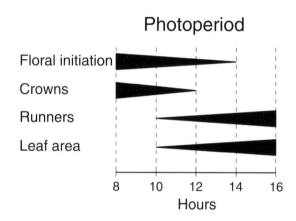

Fig. 5.7. Photoperiod regulation of strawberry plant development.

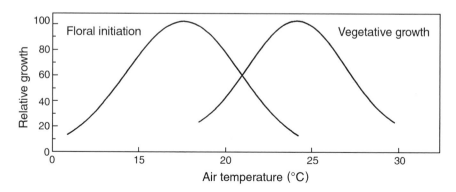

Fig. 5.8. Temperature regulation of strawberry plant development.

Strawberry development has a cyclic pattern, which is driven by the environment (Fig. 5.9). In climates with cold winters, flower buds of short-day types are normally formed in the late summer and autumn in the field, and these buds break the following spring when temperatures are warm enough for their development. The subsequent rate of floral development is strongly related to temperature (Fig. 5.10). The major part of crown branching in short-day plants occurs during the autumn when temperatures are cool and the days are short. It continues after the production of stolons has stopped due to shortness of days (10 h), and continues until a hard freeze (Darrow and Waldo, 1934).

The short-day cultivars developed for temperate climates can also be grown in mild subtropical climates (Subramanium and Iyer, 1974); although flower bud formation is restricted if temperatures get too hot (Strik, 1985). Durner *et al.* (1984) observed no floral induction under short days in plants of *F.* × *ananassa* held at 26/22°C and 30/26°C day night temperature regimes (Table 5.1). Chabot (1978) found that no flowering occurred in *F. vesca* above a 20/10°C temperature regime.

The minimum number of photoinductive cycles necessary to induce flowering in short-day strawberry plants has been variously reported to be from 7–24, depending in part on temperature (Hartmann, 1947a, b; Went,

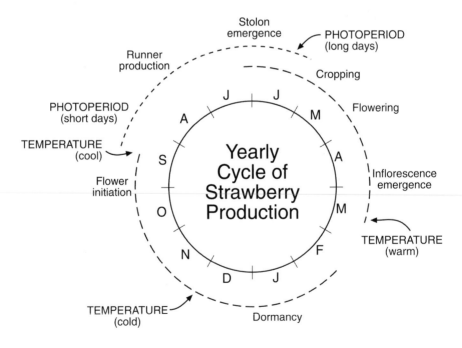

Fig. 5.9. The perennial cycle of strawberry development (reprinted by permission from Battey *et al.,* 1998).

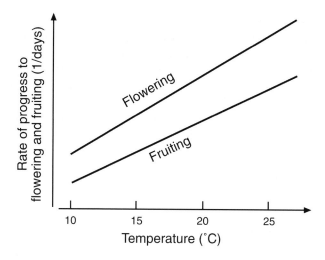

Fig. 5.10. Influence of temperature on the rate of floral and fruit development (reprinted by permission from Battey *et al.*, 1998).

Table 5.1. Average number of inflorescences and runners per plant for short-day and day-neutral strawberries grown under two photoperiods at four temperatures for 3 months (from Durner *et al.*, 1984).

Temperature (°C)	Short-day		Day-neutral	
	Inflorescences	Runners	Inflorescences	Runners
18/14	2.1a	0.0b	3.3a	1.7b
22/18	0.3b	0.0b	1.3b	2.3b
26/22	0.0b	0.8b	0.0b	2.2b
30/26	0.0b	2.4a	0.0b	3.3ab

Means followed by different letters in the same column are significantly different at $P < 0.05$.

1957; Guttridge, 1985; Larson, 1994). At higher temperatures, longer photoinduction periods appear to be needed. Ito and Saito (1962) found that under 8-h photoperiods only ten cycles were necessary for floral induction at 24°C, whereas at 30°C more than 20 cycles were required. With 16-h photoperiods, ten cycles were necessary at 9°C, but 16 were required at 17°C.

Day-neutral plants produce crowns and flower buds approximately 3 months after planting, regardless of the day length (Bringhurst and Voth, 1975; Galletta and Bringhurst, 1990). They initiate flower buds throughout the growing season, although high temperatures can inhibit bud formation as in short-day plants. Durner *et al.* (1984) found that day/night temperatures of 30/26°C almost completely inhibited flower bud initiation in the day-neutrals

'Hecker' and 'Tristar', although these cultivars were highly productive under 18/14, 22/18 and 26/22°C regimes (Table 5.1).

Although most cultivars are now categorized as day-neutrals or short-day plants, some genotypes are hard to rank precisely due to complex interactions between genotype, temperature and photoperiod (Nicoll and Galletta, 1987; Yanagi and Oda, 1993). Darrow (1966) suggested that strawberries actually range from obligate short-day to facultative short-day to complete day-neutrals. He also suggested that a range of everbearer types exist from weak, intermediate to strong. Galletta and Bringhurst (1990) have suggested that flowering in modern strawberry cultivars is regulated more by sensitivity to high temperature than photoperiod, with short-day plants being more sensitive to high temperatures than day-neutrals.

In short-day plants, stolons are produced after flowering at the base of new leaves (Fig. 5.2). They are formed most readily during long days (> 10 h) when temperatures are in the range of 21–30°C (Darrow, 1936; Heide, 1977; Durner et al., 1984; Hellman and Travis, 1988). Darrow (1936) found three times more stolons were produced under 16 h than 14 h photoperiods, and at both photoperiods, three times more stolons were formed at 21°C than at 12.8 or 15.6°C (Table 5.2). Stolons are produced until days become short in the autumn (< 10 h) and temperatures reach freezing. Runner formation in day-neutral cultivars is also highest under long days with moderate temperatures, but it is much more sporadic than in short-day types (Downs and Piringer, 1955; Smeets, 1979; Durner et al., 1984). Runner removal can stimulate and hasten branch crown development in both short-day and day-neutral types.

Leaf production in both day-neutral and short-day plants continues throughout the whole season, with the most leaves being formed under long days (Arney, 1953a, b, 1954; Nishizawa, 1990). Leaf production ceases when temperatures in the autumn fall below 0°C, and slows down greatly if

Table 5.2. Influence of temperature and photoperiod on stolon and flower production in strawberry (adapted from Darrow, 1936).

Treatments		Stolons (S) and flower panicles (F) produced per month										
		Nov		Dec		Jan		Feb		Mar		Total	
Temp. (°C)	Photo-period (h)	S	F	S	F	S	F	S	F	S	F	S	F
21	16	91	2	95	0	27	0	40	0	59	12	312	14
	14	66	5	27	0	7	1	4	10	6	24	110	40
15.5	16	57	5	15	0	0	0	0	11	0	38	71	54
	14	19	13	4	7	0	0	0	24	0	71	23	115
12.8	16	40	6	8	1	1	0	0	0	0	61	49	68
	14	18	5	2	1	0	0	0	4	0	63	20	73

temperatures exceed 30°C in the summer. Plants held at high temperatures have smaller canopies than those developing under cool conditions, and this can lead to yield reductions (Battey *et al.*, 1998; Le Mière *et al.*, 1998). Temperature optima range between 15 and 26°C, depending on cultivar (Darrow, 1936; Arney, 1953a, b, 1954; Abdelrahman, 1984). Arney (1953a) found leaf initiation in 'Royal Sovereign' to be greatly reduced above 35°C. It has also been found that plants heat treated for viruses stop growing at 35–38°C, but if temperatures are raised a few degrees a day, leaves appear undamaged for at least 6 weeks (Converse, 1987).

Roots are produced most abundantly in the spring and autumn when temperatures are cool. Strong root growth occurs at 7–32°C, but is greatest at the lower end of this range (Abdelrahman, 1984; Proebsting, 1957; Roberts and Kenworthy, 1956). The root system is the only tissue to exhibit a reduction in biomass during fruiting (May *et al.*, 1994). Roots continue to grow after leaves stop proliferating in the autumn and they remain active until the soil is frozen.

High root temperatures can greatly affect top growth, as Proebsting (1957) found that strawberry shoot growth was highest at root temperatures of 24°C and decreased sharply at higher temperatures. Brouwer (1963) and Cooper (1973) reported that shoot growth in strawberry is unaffected by root temperatures between 18 and 30°C during the growing season, but is markedly reduced at temperatures above 30°C. Nishizawa and Hori (1993) found that if the roots of plants were kept at 26°C rather than ambient temperatures during chilling, subsequent petiole and leaf growth was greatly restricted.

DORMANCY

A rest period (quiescence or ectodormacy) is induced in strawberries by short days and low temperatures (Darrow, 1936). The inductive short-day period is 4–6 weeks (Guttridge, 1985). A chilling requirement (–1 to 10°C) has to be met to break the dormancy period; temperatures around 0°C appear to be the most effective (Guttridge, 1985). Cultivars with different regional adaptations vary widely in their chilling requirements, with those from warm climates having greatly reduced needs.

Although short-day cultivars adapted to cooler climates can be grown in tropical and subtropical regions, they often require a chilling period for full productivity. Those cultivars developed for warm regions may not need one. The ability to grow well during the short days of October, November and December in North American greenhouses has classically been used as an indicator of a cultivar's regional adaptation. Those adapted to cooler regions generally grow poorly during this period and enter a rest period, whereas those adapted to warmer climates continue to grow (Darrow and Waldo, 1934).

The vigour of nursery dug stock appears to be associated with the number of chilling hours and photoperiod (Bringhurst *et al.*, 1960; Durner *et al.*, 1986; Fig. 5.11). In general, higher levels of chilling increases vegetative vigour, stimulates runner production, and concentrates ripening. Chilling enhances early season yields if it is accumulated prior to an optimal photoperiod date; the precise dates vary from cultivar to cultivar. Nursery workers manipulate the relative numbers of runners and crowns by controlling digging dates and thus exposure to chilling and photoperiod (see Chapter 6).

A new type of short-day strawberry has been developed in Israel for tropical and subtropical environments (Izsak and Izhar, 1984; Izhar and Izsak, 1995). These 'infra' short-day types do not require chilling for high yields and initiate flower bud primordia in response to longer light regimes (13.5–14 h) and higher temperatures (10–26°C) than traditional short-day plants. These genotypes are productive in the autumn and early winter when short-day types are generally incapable of bearing fruit.

COLD HARDINESS

When plants are fully hardened, crowns of the most hardy cultivars can withstand temperatures of –40 to –46°C, but those grown in warmer climates can not withstand temperatures much lower than –10°C (Harris, 1973; Marini and Boyce, 1979). Non-acclimated plants are usually killed

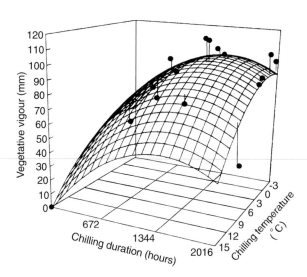

Fig. 5.11. Effect of chilling duration and temperature on vegetative vigour (increase in mean petiole length) of 'Elsanta' (reprinted by permission from Battey *et al.*, 1998).

when the crown temperature remains at −3°C for more than 1 or 2 hours. All cultivars are damaged when sharp drops in temperature occur before the plants are fully dormant in the autumn, or after warm periods in the spring have induced plant growth.

Several biochemical factors have been associated with cold hardiness. Sucrose, reducing and total sugars increase during hardening in 'Redcoat' and 'Bounty' strawberries, reaching their maximum in January (Paquin *et al.*, 1989). Proline levels also vary in these cultivars, but its role in hardiness is unclear. Percentage dry weight, starch content, total accumulation of carbohydrates and nitrogen in the roots of day-neutrals has also been associated with cold hardiness, with autumn fruit removal increasing the levels of these compounds (Gagnon *et al.*, 1990).

Three genes have been identified that have differential expression under low temperatures in strawberry (NDong *et al.*, 1997). *Fcor1* (*F*ragaria <u>cor</u>d-regulated) peaked after 2 days in leaves, crowns and roots, whereas *Fcor2* peaked after 2 weeks in the same tissues. *Fcor3* decreased within a day of exposure to cold in leaves and remained low thereafter. *Fcor1* shares homology with *blt101*, a LT-responsive gene from barley, and *ESI3*, a gene induced by salt stress in *Lophopyrum*. *Fcor3* encodes a protein similar to the spinach PSI subunit V, and PSI PsaG polypeptide of barley. The *Fcor2* protein does not show homology with any other known protein.

Acclimated plants survive sub-freezing temperatures by tolerating ice formation in crown tissues. Water migrates from cell interiors to form extracellular ice, as the temperature falls below the melting point of the cell sap (Warmund, 1993). Injury to the crowns occurs due to dehydration, resulting from extracellular freezing of the tissue water (Burke *et al.*, 1976; Levitt, 1980), and tissue separations (Warmund, 1993). Low temperature survival depends on genotype, rate of freezing, duration of low temperature, degree of temperature fluctuation and nutritional status (Boyce and Smith, 1967; Zurawicz and Stushnoff, 1977; Warmund and Ki, 1992). Straw mulch and floating row covers are commonly used to protect overwintering crowns (see Chapter 4).

Floral injury due to spring frost is strongly correlated with developmental stage and to a lesser extent genotype, although cultivar differences do exist (Hummel and Moore, 1997; Oureky and Reich, 1976). Flower buds, open flowers and young fruit are all readily injured by spring frost. Fruits are more susceptible to frost than either flowers or buds. Vegetative and flower buds have equal tolerance to frost in the spring and summer, but vegetative buds can acclimate to colder temperatures during the winter (Paquin *et al.*, 1989). Primary flowers are more susceptible to frost than secondary and tertiary ones, and styles and receptacles are more susceptible to cold than anthers (Havis, 1938; Darrow, 1966; Ki and Warmund, 1992).

Developing blooms are most sensitive just before and after they open. At this stage, temperatures only a few degrees below 0°C will cause substantial injury. Flowers in tight clusters can tolerate temperatures as low as $-5°C$. Once fruit begin to develop, temperatures around $-3°C$ can be tolerated for short periods. Sprinkler irrigation is commonly used to protect strawberries from frost injury (see Chapter 6). Ice nucleation bacteria have also been tested to raise the temperature as ice forms, but these have not proven successful on a commercial level (Anderson and Whitworth, 1993; Goulart and Demchak, 1994; Warmund and English, 1994).

PHOTOSYNTHESIS AND CARBON ALLOCATION PATTERNS

Maximum photosynthetic or CO_2 assimilation rates (A) in strawberry are comparable to many other fruit crops (Flore and Lakso, 1989). Values of A in the cultivated strawberry *F.* × *ananassa* Duch. are in the range $15–25$ μmol s^{-1} m^{-2} in the field (Hancock *et al.*, 1989a). These rates are intermediate between its progenitor species *F. virginiana* ($7–15$ μmol s^{-1} m^{-2}) and *F. chiloensis* ($20–30$ μmol s^{-1} m^{-2}), although the three species have not been compared in a common environment (Jurik, 1983; Hancock *et al.*, 1989a; Cameron and Hartley, 1990). Several factors have a significant effect on A including light levels, temperature, nutrient availability, CO_2 concentration, development, culture and propagation method.

In general, higher levels of light translate into higher A. The light saturation point in strawberries is between 800 and 1000 μE m^{-2} s^{-1} (Fig. 5.12; Cameron, 1986; Ferree and Stang, 1988). Plants grown under shade have lower maximum A than those grown under full sun, but shaded plants are more active under diffuse light (Chabot and Chabot, 1977; Chabot, 1978; Jurik *et al.*, 1982). Total light energy received during the day has a greater influence on leaf adaptation than the peak photon-flux density (Chabot *et al.*, 1979), although night interruptions increase A in everbearing types (Durner *et al.*, 1984). Plants maintained in hills have higher CO_2 assimilation rates than those grown in matted rows, presumably due to less crowding and higher light levels.

Photosynthetic temperature optima appear to vary between species and acclimation conditions. Plants of *F. vesca* held at $10/2°C$ day/night temperatures display peak rates at $15–20°C$, whereas the optimum of those held at $30/20°C$ is $25°C$. In preliminary work, Caldwell *et al.* (1990) reported that *F. virginiana* ssp. *virginiana* appeared to acclimate photosynthetically more readily to high temperature than *F. chiloensis* and *F.* × *ananassa*; but this conclusion was based on a very limited number of clones. More recently, Hancock *et al.* (unpublished) found that the mean photosynthetic temperature optima of all three of these taxa was significantly higher after maintenance at high temperature, and that heat acclimated plants showed a significantly lower reduction in A as temperatures were raised from 20 to $30°C$. In fact, some clones of *F. chiloensis* from California and commercial *F.* × *ananassa* appeared to be

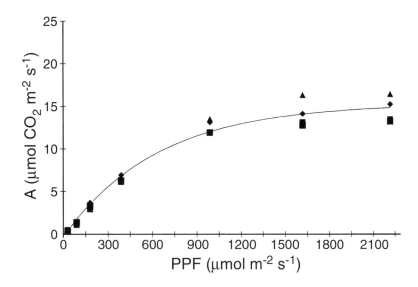

Fig. 5.12. Photosynthetic response (A) of *Fragaria* x *ananassa* to increasing light levels (photosynthetic photon flux – PPF). Measurements were made at about 375 ppm CO_2 and 25°C (redrawn from Cameron and Hartley, 1990).

better adapted to heat than *F. virginiana* from the midwestern and eastern USA, as A_{max} was significantly higher in *F. chiloensis* and *F.* × *ananassa* at the higher acclimation temperature, whereas that of *F. virginiana* decreased. Clones of *F. virginiana* from the hot, humid southern USA were not examined, so the full adaptive range of this species has still not been explored.

High root temperatures have also been shown to have a negative impact on gas exchange rates. Net *A*, stomatal conductance to water vapour and transpiration rates dropped significantly on a per leaf basis as soil temperatures were raised from 15 to 25°C, although whole plant photosynthesis increased due to greater leaf areas (Udagawa *et al.*, 1989).

Increases in fertilizer concentration often result in higher *A*, with genotypes varying in their ability to utilize increasing levels. In plants of *F. virginiana* receiving 150 ml per week of Peters 20-20-20 versus no fertilization, *A* was increased by approximately half. Moon *et al.* (1990) have found that wild clones of *F. chiloensis* vary greatly in their nitrogen use efficiency. CA11 has a greater capacity for nitrogen uptake than RCP37, but at equivalent levels of leaf nitrogen, RCP37 has higher CO_2 assimilation rates. Maximum CO_2 assimilation and carboxylation efficiency increased with increasing leaf nitrogen in both clones.

Strawberries respond positively to short term increases in CO_2 concentration. Cameron (1986) found a significant linear increase in *A* between 100 and 400 ppm. The long-term effects of increased CO_2 vary by genotype. Moon and his associates have found that 'Midway' is largely unaffected by prolonged exposure to high CO_2 levels, whereas 'Raritan' shows significant reductions in A and ribulose 1,5-bisphosphate carboxylase (Rubisco) levels. These changes appear to be at the gene transcription level, as the quantity of messenger RNA for the large subunit of Rubisco is greatly reduced in 'Raritan' after a few weeks of exposure to high CO_2.

Considerable variation occurs in the seasonal CO_2 dynamics of strawberries. The cumulative CO_2 balance of wild *F. vesca* and *F. virginiana* is negative through most of the fruiting season (Jurik, 1983). The developmental stage of strawberries has a substantial influence on CO_2 assimilation rates. Fully expanded leaves that are 10–20 days old have the highest photosynthetic rates and younger leaves are more responsive to environmental change than older ones (Jurik *et al.*, 1979). During flowering and runnering plants have higher photosynthetic rates than those during bloom (Hancock *et al.*, 1989b).

Micropropagated plants also have higher CO_2 assimilation rates than conventionally propagated plants (Cameron *et al.*, 1989). This increase can be counterbalanced by the greater shading associated with the higher densities often found in micropropagated blocks, but at similar leaf densities, tissue culture-derived plants have higher rates of photosynthesis. The enhanced A associated with micropropagation slowly dissipates over two or three generations of runnering.

Fruit removal or thinning often results in a decline in CO_2 assimilation rates on a per leaf area basis for at least a few weeks (Lenz, 1974; Choma *et al.*, 1982; Schaffer *et al.*, 1985, 1986a, b). Total photosynthetic rate on a per plant basis does not always decline, however, as more total leaf surface is produced after deblossoming (Fig. 5.13; Schaffer *et al.*, 1986a, b). Drastic leaf defoliations in excess of 66% also result in greater A per leaf surface, although whole plant levels are not completely compensated (Kerkhoff *et al.*, 1988).

Flower and runner removal can have a dramatic influence on the partitioning of photosynthate to fruit. Removal of the first year flowers in perennial cultural systems increases second year yields, whereas runner removal decreases yield (Pritts and Worden, 1988). However, excessive runnering can result in interplant competition at high densities (Wright and Sandrang, 1993). In the hill system, runner removal has been shown to significantly increase early yields of those cultivars that have a propensity for runnering (Albregts and Howard, 1986).

WATER RELATIONS

Numerous aspects of plant growth have been shown to be drought sensitive including leaf production and expansion, stolon production, root development, and berry weight and number (Renquist *et al.*, 1982a; Sruamsini and

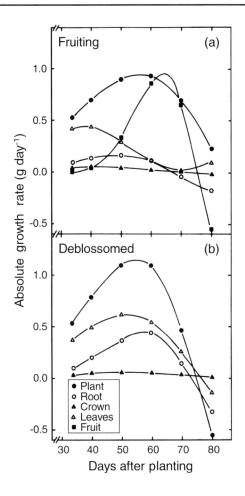

Fig. 5.13. Absolute growth rates of whole plant, root, crown, leaves and fruit of (a) normal fruiting and (b) deblossomed strawberry plants (redrawn from Forney and Breen, 1985).

Lenz, 1986; Strabbioli, 1988; Archbold and Zhang, 1991; Serrano *et al.*, 1992; Save *et al.*, 1993; Gehrmann, 1995). During active periods of growth, a dramatic mid-day drop in leaf water potential is common (Fig. 5.14).

The degree to which the various parts of the strawberry are affected by drought stress is strongly influenced by the overall stage of plant development. Leaf size is least affected during flowering (Kinnanen and Sako, 1979). Water stress imposed at the beginning of flowering has a negative impact on flower number, whereas stress imposed at later stages may actually increase flower numbers (Naumann, 1961).

Several factors impact on the drought sensitivity of strawberries.

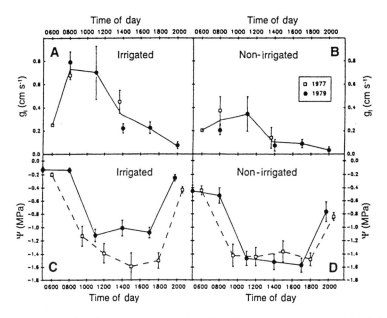

Fig. 5.14. Diurnal leaf conductance (g_l) and leaf (xylem) water potential (Ψ) for irrigated and non-irrigated first-year strawberry plantings (redrawn from Renquist *et al.*, 1982a).

Resistant strawberries osmotically adjust (Zhang and Archbold, 1993a), increase cell wall elasticity (Save *et al.*, 1993), leaf thickness (Darrow, 1966), water use efficiency (Giovanardi and Testolin, 1984) and root/shoot ratios (Hanson, 1931; Renquist *et al.*, 1982b). Drought-stressed strawberries typically have smaller leaf numbers and area (Renquist *et al.*, 1982a; Gehrmann, 1985; Archbold and Zhang, 1991), and their stomatal conductance and transpiration is reduced (Sruamsiri and Lenz, 1986). Cuticle thickness, as measured by total chloroform-extractable lipids, has been associated with levels of cuticular conductance within *F. chiloensis*, but this relationship has not held across *F. virginiana* and *F.* × *ananassa* (Archbold, 1993).

Osmotic adjustment in *Fragaria* occurs through the accumulation of carbohydrates (O'Neill, 1983; Zhang and Archbold, 1993b). In water stressed *F. chiloensis*, total soluble carbohydrate concentration increased 1.4- to 2.4-fold during wilting cycles, whereas leaf starch content dropped 4–6% (Zhang and Archbold, 1993b). Glucose and fructose were the primary carbohydrates involved in osmoregulation, accounting for more than 50% of the total osmotic potential. Total free amino acid content increased 1.8- to 2.7-fold in response to stress, with no proline being detected.

The drought tolerance of strawberries is to a large degree species specific. *Fragaria chiloensis* appears to have a greater capacity to adjust osmotically than *F. virginiana* and has greater root growth, thicker leaves and cuticles, more sunken stomata and reduced stomatal area per leaf (Darrow and Dewey, 1934; Darrow, 1966; Archbold, 1993, 1996). Leaves of *F. virginiana* typically curl and wilt after brief periods of drought, whereas leaves of *F. chiloensis* remain green and in turgor for much longer periods of stress. When potted *F. virginiana* and *F. chiloensis* clones were left unwatered, several clones of *F. chiloensis* required an average of 6–10 days before wilting, whereas one *F. virginiana* clone wilted within 2 days. However, leaf production and leaf expansion were less affected by cycles of wilting in *F. virginiana* than *F. chiloensis* (Archbold and Zhang, 1991). Clones of *F. × ananassa* performed inter-mediately to the two parental species.

Many *F. chiloensis* clones exhibit greater water use efficiency, lower leaf water potential, lower relative water content, greater membrane stability, thicker cuticles (Archbold, 1993) and greater osmotic adjustment than many *F. virginiana* clones, although there is considerable variability within both species (Archbold and Zhang, 1991; Zhang and Archbold, 1993a; McDonald *et al.*, 1994). In a comparison of single clones of *F. virginiana* and *F. chiloensis*, *F. chiloensis* wilted at a leaf water potential 1.1 mPa lower, a leaf osmotic potential 0.45 mPa lower and a relative water content 14% lower than *F. virginiana* (Zhang and Archbold, 1993b). Cultivated *F. × ananassa* appear to be intermediate between the two octoploid species in drought tolerance, although cultivars exist which appear to adjust osmotically as well as some *F. chiloensis* (Darrow and Sherwood, 1931; Renquist *et al.*, 1982a; Save *et al.*, 1993; Zhang and Archibold, 1993a, b).

NUTRITION

Strawberries grow and produce satisfactorily in a wide range of soil types from sands to heavy loams. They are also tolerant of a wide range of soil pH values, but they grow and produce best on soils with a pH of 6.0– 6.5. Highest yields are obtained when plants are grown in deep fertile soil, with high organic matter and good drainage.

As with all other plants, the strawberry has critical requirements for a number of nutrients (Marschner, 1986; May and Pritts, 1990a; Maas, 1998). Adequate levels of nitrogen (N), phosphorus (P) and potassium (K) are essential for proper growth and development (Ruef and Richey, 1925; Boyce and Matlock, 1966; John *et al.*, 1975). Excessive levels of N can lead to soft fruit, delayed ripening, lower yields, and increased powdery mildew and mite pressure (Stadelbacker, 1963; Voth *et al.*, 1967; May and Pritts, 1990a). Calcium (Ca) levels are important determinants of fruit firmness (Eaves and Leefe, 1962). Boron (B) deficiency has been shown to reduce viable pollen production, pollen germination and receptacle expansion (Guttridge and Turnbull, 1975; Neilson and Eaton, 1983), and can result in the formation of small, fasciated fruit and reduced primary and lateral root growth (Ulrich *et*

Table 5.3. Leaf deficiency symptoms in strawberries (sources: Johanson, 1980; Pritts, 1998).

Nutrient	Symptoms
Boron	Early leaves develop a squared off appearance. Later emerging leaves are stunted and twisted, with the small leaflets being asymmetrical in size
Calcium	Begins as a tip-burn in the runner tips and folded leaves emerging from the crown. Fully developed leaves are crinkled and a necrotic band appears across their centres. Fruit are dull coloured and soft
Iron	Symptoms first appear in young leaves as yellow interveinal areas inside green veins. Emerging leaves become progressively smaller and eventually are almost white
Magnesium	Pot studies have shown a reddish to purplish coloration between veins extending toward the midrib, but in field observations only necrosis has been observed
Manganese	Dull, interveinal chlorosis develops that does not extend into marginal serrations. Veins remain greener, but not as sharply distinctive as in iron. A pin-point purple stippling eventually appears on some cultivars, next to a green marginal halo
Nitrogen	Starts as undersized, yellowish-green foliage. Tips of older leaves become red with the colour gradually spreading inward until whole leaves are brilliant orange–red
Phosphorus	Starts as stunted, dark-green foliage with bluish-purple coloration. The bluish colour begins in small veins, gradually spreads to whole veins and ultimately colours the whole leaf
Potassium	In pot studies, margins become necrotic or reddish from the leaf tips inward, but the basal portions remain green, causing a triangular 'Christmas tree' pattern. In the field, the red colour is rarely observed, only necrosis
Sulphur	Similar to nitrogen deficiency. Margins of older leaves develop brownish-black tips. Some red speckling and yellowing is observed in the field

al., 1980; Johanson, 1980). Zinc (Zn)-deficient strawberries have smaller leaves and fruit, and reduced yields (Ulrich et al., 1980). Iron (Fe) deficiencies lead to reduced vigour and chlorotic leaves.

Some nutrients are more limiting than others (Pritts, 1998). N and K supplementation is almost always required. Preplant incorporation of P is sometimes required, but usually not after planting. Deficiencies in Ca are often observed, but are usually due to factors that limit mass flow such as low soil moisture or cool, cloudy, humid weather rather than low Ca in the soil. Boron is also frequently low in strawberry plantings due to leaching, and can be easily brought to excess by over application. Iron, magnesium and manganese are usually only deficient in producers' fields when pH is excessively high or the soils have a high lime content (King et al., 1950; Berger, 1962; Renquist and Hughes, 1985).

Table 5.4. Sufficiency ranges for foliar nutrient levels in strawberry in mid-summer (from Pritts and Handley, 1998).

Nutrient	Deficient below	Sufficient	Excess
N (%)	1.9	2.0–2.8	4.0
P (%)	0.20	0.25–0.40	0.50
K (%)	1.3	1.5–2.5	3.5
Ca (%)	0.5	0.7–1.7	2.0
Mg (%)	0.25	0.3–0.5	0.8
S (%)	0.35	0.4–0.6	0.8
B (ppm)	23	30–70	90
Fe (ppm)	40	60–250	350
Mn (ppm)	35	50–200	350
Cu (ppm)	3	6–20	30
Zn (ppm)	10	20–50	80

Soil nutrient interactions can play an important role in the regulation of strawberry productivity (May and Pritts, 1990a). Kirsch (1959) found significant interactions between soil levels of N, P, K and lime. Phosphorus increased marketable yields, but only when N and lime were not applied. K only increased yields when lime was added. Zurawicz and Stushnoff (1977) found that the plants most resistant to cold stress had the highest P:K ratio. May and Pritts (1990b) found reduced yields associated with extremes in soil phosphorus to boron (P:B) ratios, with fruit number per truss being the most greatly affected yield component. These P–B interactions only occurred at high pH (6.5).

Deficiency symptoms have been described in pot-grown strawberry plants for all the major elements (Table 5.3; Plates 5–10). However, considerable differences in cultivars often exist in the field, and the environment can play a critical role in symptom expression. In addition, multiple deficiencies often occur together, such as Fe, Mn, Zn and Cu at high pH. Deficiencies are best detected by conducting a leaf analysis before the plants exhibit symptoms.

Foliar sufficiency ranges for most of the critical elements have been proposed (Table 5.4), together with soil test recommendations. Leaf nutrient status is currently assessed after harvest when foliar levels are relatively stable, although other periods of fluctuating levels may in some instances give a more accurate representation of nutrient needs (May and Pritts, 1994). Most nutrient deficiencies are elevated by soil applications of fertilizers or the injection of minerals through irrigation systems (see Chapter 7). Manure and composts are also used occasionally (Pritts and Handley, 1998). Strawberries are mycorrhizal, and inoculations of these beneficial fungi can increase P content and dry weight accumulation in strawberries, particularly at low P levels and during the reproductive phase, although this practice is rarely done commercially (Holevas, 1966; Hughes *et al.*, 1978; Dunne and Fitter, 1989; Khanizadeh *et al.*, 1995).

CONCLUSIONS

The strawberry is a herbaceous perennial that is composed of several different meristems, whose development is regulated by the interaction between photoperiod and temperature. The strawberry plant has a central crown from which leaves and roots emerge. At the top of each leaf is an axillary bud, which can produce runners or inflorescences depending on environmental conditions. In short-day cultivars, inflorescences are generated when days are short and temperatures are cool, whereas stolons are produced under long days and warm temperatures. The minimum number of photoinduction cycles ranges from 7 to 24 days, depending on genotype. In day-neutral cultivars, flowers are produced in a cyclical pattern regardless of day length as long as temperatures remain relatively cool.

A rest period is induced in strawberries by short days and low temperatures. Cultivars vary widely in their chilling requirements, but most need at least some chilling to be fully productive. There is also a wide range in cold hardiness, depending on where a cultivar was bred. Acclimated plants survive sub-freezing temperatures primarily by tolerating ice formation in crown tissues. The tolerance of flowers to frost also varies, but to date, most differences have been associated with later bloom dates rather than bud tolerance *per se*. It seems likely that breeders could further improve bud tolerance by transferring genes from native species or engineering resistance with cloned cold tolerance genes.

Photosynthetic rates in strawberries are comparable to most other fruit crops. Maximum CO_2 assimilation rates can be in excess of 30 μmol m^{-2} s^{-1}, and light saturation points fall around 800–1000 μE m^{-2} s^{-1}. Fruit are a very strong sink in strawberries, and more total leaf surface is produced after deblossoming. Photosynthetic temperature optima seem to vary widely among native species and genotypes, and this could prove to be useful in breeding types better adapted to temperature extremes.

Most strawberry cultivars are highly subject to drought, although there is a substantial amount of variability in tolerance among species material that could be used in breeding programmes. Resistance to drought comes in many forms including osmotic adjustments, increases in leaf thickness, higher water use efficiency and increased root/shoot ratios.

A considerable amount of work has been conducted to determine the optimal nutrient requirements of strawberries. In field situations, the elements most frequently limiting are N, K, B and Ca, whereas P, Fe, Mg and Mn are more rarely a problem. Although deficiency symptoms have been described for most of the major elements, these are highly subject to cultivar and environmental variation. The most accurate way to adjust fertilization practices is through foliar analysis before the deficiencies appear. Sufficiency levels have been proposed for strawberry for most of the important macro- and micronutrients, although these have not all been rigorously established.

6

CULTURAL SYSTEMS

INTRODUCTION

Strawberry yields and fruit quality are greatly influenced by a complex set of interactions between genotype and environment. Important environmental factors include temperature, photoperiod, length of rest period, winter hardiness, disease resistance, tolerance to various soil conditions and high temperature resistance. The response of individual genotypes can vary dramatically depending on the array of conditions surrounding it. As a result, numerous different cultural systems have been developed across the world to exploit the particular conditions found in individual regions.

As mentioned briefly in Chapter 1, there are two major production systems utilized – hills and matted rows (Table 6.1; Plates 1 and 2). The matted row system employs runners as the primary yield component. Both mother and daughter plants are allowed to runner freely, with periodic training into narrow rows. The hill or 'plasticulture system' relies on crowns as the primary yield component, and any runners that form are removed.

The hill system is used to grow day-neutral varieties everywhere, and short-day types in areas that have warm winters and either hot or moderate

Table 6.1. Characteristics of the two major strawberry cultural systems.

Characteristic	Hill	Matted row
Winter temperatures	Mild	Cold
Planting dates	Summer or winter	Spring
Planting distance (between plants)	20–30 cm	45–60 cm
Bed height	Raised	Flat
Mulch	Clear or black plastic	Straw
Irrigation	Trickle	Overhead
Production seasons	One to two	Three to five

summers such as California, Florida, Italy and Spain. Matted rows are used to grow short-day cultivars in climates with short summers and cold winters such as continental Europe and northern North America (Hancock *et al.*, 1998). In general, cultivars perform best in one cultural system or the other, although some are more flexible than others (Hancock *et al.*, 1983, 1984a, b). Annual systems can generate yields in excess of 30 t, whereas well maintained matted rows rarely exceed more than 10 t.

Matted row and annual hill systems have been modified in a number of ways to fit the specific requirements of individual producing regions. Many factors are varied across the world, including planting date, the chilling history of planting stock, mulch, bed height, irrigation, nutrition, weed management and renovation. This chapter, summarizes these cultural variations and contrasts them. Open versus closed systems of production are also described and the various methods of plant propagation and harvesting are outlined.

HILL AND MATTED ROW CULTURE

Planting dates

Strawberries are planted at several different times of the year, depending on variety, climate, location and the training system employed (Galletta and Bringhurst, 1990; Larson, 1994). Winter and autumn temperatures play an important role in selecting planting dates, together with whether the plants are short-day or day-neutral types.

Day-neutral types produce fruit approximately 3 months after planting, regardless of planting date, although they perform poorly when temperatures exceed 26°C (see Chapter 5). They have become very popular in California as a means of extending the traditional short-day harvest season. In central coast areas, plants are most often set in the field in the late autumn from 15 October to 5 November, whereas in southern California they are commonly planted in the late summer from 15 August to 15 September (Strand, 1994). Day-neutral types are far less popular in regions with cold winters, but are sometimes planted in the late summer in Europe to obtain a late crop, or in the spring in North America to extend the fruiting season into the late summer. Fruit production in day-neutrals is cyclic, with periods of flowering and fruiting interspersed. The first set of flowers is often removed to enhance plant vigour and subsequent yields.

Short-day plants grown in matted rows are usually set in the early spring as soon as the soil can be worked. First year flowers are commonly removed to maximize plant vigour and runner production (Hancock and Cameron, 1986). Short-day types grown in hills are planted in either the summer or late autumn/early winter, depending on climate and the desired production season. In regions with relatively mild winters (lows of 0–10°C) and moderate summers (highs below 30°C), strawberries are often planted in the summer to be cropped the following spring. Typical summer planting dates range from early August to mid-September (Strand, 1994; Poling, 1994). In areas with very mild winters and hot summers, plants can also be set in the late autumn for a winter crop. Late autumn planting dates range from late September to 30 November (Strand, 1994; Maynard *et al.*, 1996), depending on specific location. Any runners that form in hill systems are removed, but runner production is minimized by using plants pre-acclimated to produce flowers rather than runners. In addition, the short day lengths of the late autumn and winter months discourage runner production (see Chapter 5).

The variations in planting date optimize several different climatic–physiological relationships (Larson, 1994). Spring plantings are used in conjunction with matted rows in regions where the autumn is too short for the formation of many inflorescences and crowns and high yields depend on the production of stolons in the summer (Hancock *et al.*, 1998). The summer planting system is generally used with hills where temperatures remain above freezing long enough for ample crown and flower bud production, but the winter is too cold for continued growth. The late autumn planting system is used with hills in those regions with winter temperatures that are high enough to sustain growth. This system produces earlier fruit with higher quality and size than summer plantings, but has lower yields and is less dependable (Galletta and Bringhurst, 1990).

Very precise local recommendations have been developed for when planting stock should be dug and set in the annual systems, as productivity and harvest season are closely related to chilling hours (Fig. 6.1; Voth and Bringhurst, 1990; Faby, 1997). In the late autumn/ winter planting system, both partially chilled and fresh dug plants are used, depending on the climate

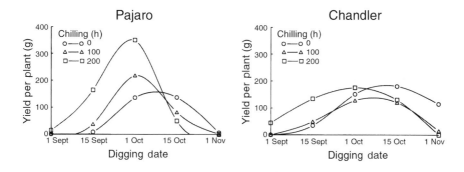

Fig. 6.1. Yield response (g per plant) of two strawberry cultivars to different photoperiod and chilling treatments prior to planting in a Florida winter production system (redrawn from Durner and Poling, 1988).

and desired harvest dates (Durner *et al.*, 1986; Voth and Bringhurst, 1990). Plants with partial chilling (300–500 hours below 7.2°C) produce flowers a few months after planting, but make little vegetative growth and produce few runners (Galletta and Bringhurst, 1990). Freshly dug plants produce fruit immediately from nursery-initiated flowers and then later produce another round of flowers. In the summer planting system, 'frigo' plants are commonly employed, although freshly rooted 'plug plants' are used in the southern and mid-Atlantic parts of the USA. Planting stock for matted row systems generally consists of cold-stored, 'frigo' plants which have had their chilling requirements satisfied (see below).

Mulch

Two main kinds of mulch are utilized in strawberry cultivation, straw and plastic. Biodegradable paper mulches have shown promise as substitutes to plastic, but they have not been widely utilized (Albregts and Howard, 1972). Straw is used in conjunction with matted row systems. It is placed next to the rows in the springtime in areas with mild winters to keep the fruit clean, and it is placed on top of the plants in late autumn/early winter in those areas with cold winters (Plate 11) and moved to the alleys in spring. The mulch protects plants from the cold by reducing the variations in soil temperatures that cause heaving, and as a result minimizes root breakage.

Clear, black and occasionally white polyethylene mulches are used in the annual hill systems to keep fruit clean and modify soil temperatures for maximized growth and development (Renquist *et al.*, 1982b; Fear and Nonnecke, 1989; Voth and Bringhurst, 1990). Numerous other colours have also been tested, but they are not widely employed. The primary use of plastic mulches is to regulate soil temperature. The amount of radiant energy penetrating through the mulch regulates soil temperature; it is highest for clear film and lowest for white over black (Table 6.2).

Black polyethylene is utilized in areas where summer temperatures are prohibitive for fruit development and winter temperatures are high enough to allow for good root growth. Black plastic can be particularly valuable in controlling weeds. Transparent plastic is used in climates where winter temperatures are not sufficient to maximize root growth and summer temperatures are not prohibitive. White mulch also increases soil temperature, but not as much as clear plastic, and provides a slightly later harvest than clear plastic (Strand, 1994).

Spunbonded fabric covers and light-weight polyethylene films are sometimes used in the cooler production areas as a means of hastening fruit ripening and protecting against winter cold (Fig. 6.2; Pritts *et al.*, 1989; Austin, 1991; Gast and Pollard, 1991; Hochmuth *et al.*, 1993). The covers are most commonly used in annual hill systems, where they are placed over the

Table 6.2. Effect of various polyethylene bed mulches on bed temperatures at 7.6 cm soil depth at Irvine, California. Mean separation by Duncan's multiple range test, $P = 0.05$ (adapted from Voth and Bringhurst, 1990).

| | Mean bed temperature (°C) | | | |
| | January–February | | March | |
Mulch	a.m.	p.m.	p.m.	Overall mean
Clear	8.9	25.6	31.1	21.8a
Brown	8.3	22.2	26.1	18.9b
None	7.5	22.7	23.9	18.0b
Black	6.1	21.6	22.7	16.8c
White	5.0	17.8	20.0	14.3d
White/black	5.5	16.6	19.5	13.8

Mean followed by different letters in the same column are significantly different at $P < 0.05$.

Fig. 6.2. A strawberry field in the UK covered with a floating row cover (picture a gift of David Simpson).

rows in the late autumn, and are kept in place until bloom. Timing of placement and removal can be critical, as placement and removal too early can reduce yields (Pritts *et al.*, 1989; Gent, 1990). Although research results have been variable and dependent on duration of snow cover, mulch application appears to be most beneficial if it is placed in the field a couple of weeks before plant development normally ceases in the autumn, and is removed a week or so after normal development begins in the spring. Harvests can be made up to a week earlier using this system and the fabrics can act as effective winter mulches.

Various living mulches have also been tested in strawberries for wind protection and weed control. Whitworth (1995) showed that residues of rye and wheat tended to suppress the growth of weeds and increase strawberry growth in pot and field studies. Likewise, Smeda and Putnam (1988) found that rye and wheat provided good early weed control, without having a negative impact on strawberry yields. Newenhouse and Dana (1989) found that living mulches of grasses, particularly ryegrass (*Lolium perenne* cv. 'Regal') helped to protect plants during bloom, and discouraged weed growth between rows. Pritts and Kelley (1993) found that interseeding sudangrass after harvest and mowing it twice a year provided acceptable weed control without herbicide.

Bed height and spacing

Plants are grown on flat or raised beds. Raised beds are usually 15–25 cm high and wide enough to accommodate two or four rows of plants (Fig. 6.3; Plate 2). The raised beds lose more water than flat beds so are typically covered with

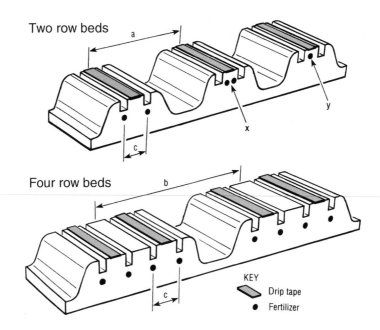

Fig. 6.3. Raised strawberry beds with two or four rows of plants (reprinted with permission from Strand, 1994). a, spacing for two-row beds is 100–130 cm; b, spacing for four-row beds is 150–170 cm; c, plant spacing is 30–35 cm.

plastic mulch. They are more subject to winter damage due to plant heaving, but they provide better soil aeration, drainage and are easier to harvest. Matted rows are generally grown on flat beds, as runner training via cultivation is very difficult on ridged surfaces. Hills can be successfully grown in both types of bed, but are usually planted on ridges.

Plants are typically set 20–30 cm apart within hill rows, and 45–60 cm apart in matted rows. The goal in both systems is to maximize the number of crowns per surface area, without reducing yield and fruit size due to crowding (Fig. 6.4). Optimal spacings are dependent on both climatic conditions and the vigour of individual cultivars.

Irrigation

As with all other crops, strawberries generally require irrigation for optimal growth and yield (Dwyer *et al.*, 1987; Strabbioli, 1988; Mannini and Anconelli, 1993; Archbold, 1996). Irrigation can come from overhead sprinklers, drip, furrows or a combination of sprinklers and drip. Drip irrigation is the dominant mode of water placement used commercially, due to its efficiency, but overhead irrigation remains popular in cold regions where spring frost control is required (Pritts and Handley, 1998), and is used with all planting systems during plant establishment. Double rows of ground level tape are typically utilized in drip systems with emitters near each plant (Fig. 6.3). Furrows are now rare except in areas where the cost of irrigation pipe and tape is prohibitive.

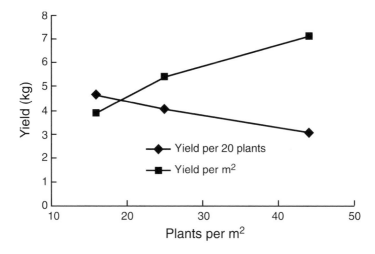

Fig. 6.4. The effects of planting density on yield of 'Elsanta' (reprinted with permission from Battey *et al.*, 1998).

Water application rates vary greatly depending on climate and cultural system (Galletta and Bringhurst, 1990 ; Archbold and Zhang, 1991). During the 2-week establishment period with overhead irrigation, water use is from 175 to 300 mm ha^{-1} in California (K. Larson, 1999, California, personal communication), to 1200 mm ha^{-1} in Florida (Clark *et al.*, 1996; El-Farhan and Pritts, 1997). Strawberries grown in Mediterranean climates on plastic-covered beds with drip irrigation are given 1200–2000 mm of water per season, depending on location. Precise water use needs have not been calculated for matted row systems grown under overhead irrigation, but general recommendations are for about 60 mm of water per week throughout the growing season, depending on evaporative demand (Pritts and Handley, 1998).

Irrigation scheduling is typically based on soil moisture content in the top 30 cm of the root profile where most roots are located (El-Farhan and Pritts, 1997; Pritts and Handley, 1998). Tensiometers or gypsum blocks are frequently used so the grower can keep soil moisture content above 50% available water (Renquist *et al.*, 1982a, b; Dwyer *et al.*, 1987; Strabbioli, 1988; Clark *et al.*, 1996). Scheduling also can be based on crop coefficients that reflect atmospheric evaporative demands or vapour pressure deficits that vary with increasing leaf area as the season progresses (McNiesh *et al.*, 1985; Strabbioli, 1988; Clark *et al.*, 1996). McNiesh *et al.* (1985) proposed that the crop coefficient for strawberries is $0.55 < K_c < 0.80$.

It has been suggested that predawn measurements of leaf water potential might be used to schedule irrigation (Archbold and Zhang, 1991). For example, Dwyer *et al.* (1987) used a minimum potential value of −0.25 mPa to initiate irrigation, rather than 50% soil moisture content or 2.5 cm of rain a day. However, water potentials are dependent not only on water status, but also on cultivar and stress history (Archbold and Zhang, 1991) and low predawn water potentials may predispose a plant to midday stress as temperatures and vapour pressure deficits rise (Johnson and Ferrell, 1983). Leaf guttation has been suggested as a visual means to estimate adequate leaf water potential (Glenn and Takeda, 1989), although guttation ceases at −0.11 mPa, a level that may be higher than those significantly affecting plant processes (Archbold and Zhang, 1991).

Overhead irrigation is commonly utilized for frost protection in areas prone to spring frosts (Pritts and Handley, 1998). Blossoms are protected by continuously covering them with small quantities of water that freezes and releases latent energy. Most commonly, nozzles are used that deliver about 25 mm h^{-1}. When there are open blossoms, irrigation is usually begun when temperatures approach 0°C during the night and is continued until temperatures rise above freezing the following day. At earlier stages of development, the blossoms can tolerate lower temperatures without sustaining injury (see Chapter 5).

Nutrition

Fertilization practices vary widely across geographical regions and cultural systems. In general, nitrogen is applied most frequently and in the greatest amount, followed by potassium and phosphorus. Soil and foliar analysis are generally recommended for precise application rates of these and other nutrients. Typical N fertilization rates are in the range 110–450 kg ha^{-1} (Kolb, 1986), with the lower rates being used in perennial systems and the higher rates in annual systems (May and Pritts, 1990a). Although it is not commercially utilized, leaching of nitrates can be prevented by maintaining soil N as ammonium through chemical inhibition of nitrification (Bunday and Bremner, 1974; Huber *et al.*, 1977) and denitrification (McElhannon and Mills, 1981; Mills and Pokorny, 1985).

In matted row systems, 30–45 kg N ha^{-1} in a non-coated fertilizer is generally side dressed within a few weeks of planting (Pritts and Handley, 1998). In subsequent years, 80–100 kg N ha^{-1} is usually applied after harvest. Additional applications of N are sometimes applied in autumn or spring, if conditions warrant it. However, care is generally taken with spring applications, as they encourage excessive vegetative vigour and increased disease problems.

Uniform N levels are maintained in hill systems using controlled release fertilizers (Voth *et al.*, 1967; Patel and Sharma, 1977; Voth and Bringhurst, 1990) and fertigation (Hochmuth and Albregts, 1994). In California, N at 200–270 kg ha^{-1} is typically placed in the planting slot or banded adjacent to the rows. A typical analysis would be 22-7-20 or 18-8-13 N-P-K, depending on location, cultivar and meteorological events. In addition, some growers might also apply as much as 110 kg ha^{-1} of N through the drip system at regular intervals during the growing season (K. Larson, 1990, California, personal communication). In Florida, only about 35 kg ha^{-1} of N and K are applied before planting, and most fertilization is done through the drip system (Table 6.3). Micronutrients are applied as necessary.

Table 6.3. Nitrogen and potassium fertigation recommendations for strawberries (adapted from Hochmuth and Albregts, 1994).

Stage in season	Injection rate (kg ha^{-1} day^{-1})	
	N	K$_2$O
First 2 weeks	0.4	0.4
February/March	1.0	1.0
All other months	0.8	0.8

Weed control

A wide range of broadleaf weeds and grasses cause problems in all production areas (Galletta and Bringhurst, 1990; Hemphill, 1990; Strand, 1994). There are significant regional differences in weed populations, but some of the most important perennial weed pests are Bermuda grass (*Cynodon dactylon*), bindweed (*Convolvulus arvensis*), Canada thistle (*Cirsium arvense*), hedge bindweed (*Convolvulus sepium*), horse nettle (*Solanum carolinense*), horsetail (*Conyza canadensis*), quackgrass (*Agropyron repens*), red sorrel (*Rumex acetosella*) and yellow nutsedge (*Cyperus esculentus*). Some of the most troublesome annual broadleaves are little annual sowthistle (*Sonchus oleraceus*), burclover (*Medicago polymorpha*), cheat or chess (*Bromus* spp.), chickweed (*Stellaria media*), common groundsel (*Senecio vulgaris*), filaree (*Erodium* spp.), lambsquarters (*Chenopodium album*), mallow (*Malvia parviflora*), pepper grass (*Lepidium* spp), prickly lettuce (*Lactuca scariola*), purslane (*Portulacca* spp.), ragweed (*Ambrosia* spp.) and sweet clover (*Melilotus* spp.). Some of the most noxious annual grasses are barnyard grass (*Echinochloa cruz-galli*), bluegrass (*Poa annua*), crabgrass (*Digitaria* spp.), autumn panicum (*Panicum dichotomiflorum*), foxtails (*Setaria* spp.) and wild barley (*Hordeum leporinum*).

In hill culture, growers have traditionally relied on annual pre-plant fumigation and mulches for weed control (Voth and Bringhurst, 1990; Strand, 1994). The whole field is commonly covered with a tarpaulin, and a broad-spectrum soil fumigant, such as a combination of methyl bromide and chloropicrin is injected under it. Pre-plant fumigation is also used by some matted row growers to initially control weeds and soil pathogens, but weed pressure steadily increases as the fields age. Unfortunately, there are only a few herbicides available that can be used in the establishment year, and multiple applications of herbicide are generally needed each year to control weed populations. This is of particular concern now for matted row growers, but will also become important to growers using annual hills, when methyl bromide is restricted (see Chapter 8).

Several different classes of herbicides are available to strawberry growers (Hemphill, 1990). There are pre-emergent herbicides that are soil-active and are applied to the bare ground before weeds emerge. Examples are napropamide and DCPA (dimethyl 2,3,5,6-tetrachloro-1, 4-benzenedicarboxylate) which control a number of annuals and some broad-leaved weeds. There are also post-emergent herbicides that are foliar active and are applied after the weeds emerge as an over-the-top spray. Some of these are selective and are effective against only certain kinds of weeds, such as sethoxydim and fluazifop-butyl, which control actively growing grasses, and 2,4-D (dichlorophenoxyacetic acid), which is used to control broadleaf weeds after renovation. Other post-emergent herbicides are non-selective and kill any green tissue they touch. Examples are glyphosate which moves systemically throughout the plant, and paraquat which is a contact herbicide which is not translocated. Some herbicides control both pre- and postemergence weeds such as terbacil, which is active against a number of broadleaf weeds. All of these herbicides have restrictions on where they can be used and when.

Several non-chemical techniques can help to control weed populations. The use of living mulches in weed control has already been discussed. The strawberries can even be planted directly into a killed sod. In addition, pre-plant cover crops such as marigold and sudangrass can be seeded in a field to reduce germination of weed seeds and ploughed down when the desired amount of growth has been achieved (Pritts and Kelly, 1993). Leaving fields fallow with repeated tillage before planting can also be effective, as weed seeds are continually exposed and allowed to germinate.

Renovation

Although renovation is not an issue in annual hill systems, except for runner removal in any beds held over, renovation is thought by many to be key to long-term productivity in matted row systems (Pritts *et al.*, 1992). Typically, renovation is initiated immediately after harvest with an application of the herbicide 2,4-D to control weeds (Nylund, 1950). This must be done before blossom development begins at about 1 August, or misshapen fruit will be produced the next year. About a week later, leaves are removed with a mower before fertilizer is spread. Typically rows are narrowed with a tiller to 20–25 cm, and plant densities are sometimes thinned to 10–15 cm apart (Galletta and Bringhurst, 1990).

Experimental tests indicate that these renovation practices can have a positive effect on yield and fruit size, but just as commonly, they have little impact. For example, some studies have indicated that leaf removal enhances yield (Waldo, 1939; Guttridge and Mason, 1966; Mason, 1966; Nestby, 1985), whereas several other studies have shown no effect of leaf removal (Wilson and Rogers, 1954; Hughes, 1972; Pelofske and Martin, 1982) or negative effects if plants are stressed (Pritts, 1988). Narrowing row widths increases efficiency per bearing surface, but has only rarely been shown to increase yields (Crane and Haut, 1939; Craig *et al.*, 1983), unless row centres are spaced closer together (Hill and Haut, 1949; Buckley and Moore, 1982). Although a number of studies have demonstrated that very high densities are negatively associated with yield and fruit weight in established plots (Hill and Haut, 1949; Craig *et al.*, 1983; Hancock *et al.*, 1984b), there are few studies that document increases in fruit size or total yield after thinning (Popenoe and Swartz, 1985; Pritts *et al.*, 1992). Plant vigour, the level of crowding and pest pressure may dictate whether these practices are beneficial.

Modified training systems

In some instances, growers have staggered summer field planting dates to obtain continuous production (Beech *et al.*, 1986; Baumann and Daubeny, 1989; Lieten and Baets, 1991). In this system, called 'waiting beds', plants are set in the field at low densities as early as possible, deblossomed and allowed to develop multiple crowns (Chercuitte *et al.*, 1991). The plants are then cold stored and are set every 1–2 weeks in the field the following year. Yields from this system are low, but are above the economic threshold point due to expanded markets.

Three other variations of the annual hill and matted row systems are employed to a limited extent: (i) spaced matted rows, (ii) solid set matted rows, and (iii) ribbons. In spaced matted rows, plants are planted at the same spacing as traditional matted rows, but they are thinned to lower densities through periodic runner removal. Densities are generally kept to 8–10 cm between plants. This is a compromise system that produces slightly larger fruit than the traditional matted row at slightly higher planting and training costs. In the solid set system fields are planted at high densities without rows (Ricketson, 1968). This system is used to maximize fruit production for mechanical harvesting, where fruit numbers are much more important than fruit size (Dale and Hergert, 1990).

In the ribbon row system, mother plants are set 5–10 cm apart in close spaced rows (90 cm), and all runners are removed. It is felt that by planting high densities of large-crowned plants yields will be increased enough to compensate for the higher initial high planting costs (Scheel, 1982). Ribbons have been shown experimentally to give higher yields and larger fruit in the first full production year, but this advantage is lost in subsequent years (Hancock, 1987).

TUNNELS AND FORCING SYSTEMS

Most strawberries are grown in the open, but plastic tunnels and greenhouses have become popular for season extension and off-season production in parts of Europe, Korea and Japan, particularly in areas with mild winters (Plates 12–14). They are frequently used to hasten ripening by 3–4 weeks, but they can also be used to produce autumn and winter crops. They are also valuable in rain protection. High percentages of malformed, small fruit are frequently produced in tunnels (Lopez-Galarza *et al.*, 1993), but this problem can be alleviated by improving pollination.

By combining open culture with tunnels and greenhouses, virtually year-round production can be accomplished even in regions with relatively cold winters. For example, in the UK, a combination of glasshouse, plastic tunnels and field culture is used to produce strawberries throughout most of the year. Mostly short-day plants are cultured, although day-neutrals are used for late summer and autumn production (Table 6.4). In Belgium, greenhouses are used to produce fruit in the early spring, late summer and late autumn/early winter, with tunnels being used to fill all the other open periods except late winter (Lieten and Baets, 1991). Most producers set plants directly in the soil, but in Belgium and The Netherlands, plants are more commonly set above the soil in buckets or bags (Fig. 6.5). In Japan, 60 ha of strawberries are grown in walls built of cement blocks (Oda and Kawata, 1993). Some plants are also grown hydroponically (Cooper, 1979), but bags and buckets are the most common soilless systems.

Intense annual cultural conditions are employed in the soil systems, including preplant fumigation, drip irrigation, plastic mulches and high plant densities (Rosati, 1991). Supplemental heating, cooling and lighting are also

Table 6.4. Methods of season long extension in the UK.

Harvest period	Cultural technique
March/April	Glasshouses
April/May	Multiple tunnels
May/June	Simple tunnels (French)
June/July	Traditional open field
July/August	Waiting bed plants in open field
August/October	Day-neutrals
September/October	Plastic houses
October/November	Glasshouses

Fig. 6.5. Strawberries growing in bags in an English greenhouse (picture a gift of David Simpson).

commonly employed (Fig. 6.6). Twin hill rows are often used on raised beds with plants being separated by 20–35 cm. Very large crowned plants are produced for planting in 'waiting' beds or high elevation nurseries. Gibberellic acid (GA) is sometimes used to enhance earliness and elongate fruit clusters; CO_2 is also sometimes added to increase productivity (Oda, 1997). Harvesting usually begins 7–8 weeks after planting.

In the soilless systems, the bags and buckets are filled with a combination of mostly peat moss and styromull or perlite (Lieten and Baets, 1991; Lieten, 1993). The bags are generally shaped about 40 × 25 × 10 cm and contain 8–10 l of substrate. There are small perforations in the bottom of the bags for drainage. Normally, four plants are set in each bag, resulting in plant densities of 12 plants m^{-2} or 48,500 plants per acre (120,000 plants ha^{-1}). Water and nutrients are supplied by one drip emitter per bag with a flow rate of 2–3 l h^{-1}. CO_2 enrichment has been shown to be beneficial (Lieten, 1997).

Multi-crowned waiting bed plants are usually used in this system (Lieten and Baets, 1991). Year-round production is possible by holding dormant plants in cold storage and planting them at the appropriate stage in heated greenhouses and tunnels. Nursery plants are generally set in August at 25 cm spacing in rows 35 cm apart. They are dug when dormant and stored slightly below freezing until planting.

HARVESTING

Hand picking

Most strawberry producers pick their own fruit for local and retail markets, although consumer-harvested fruit ('U-pick') is popular in some areas, par-

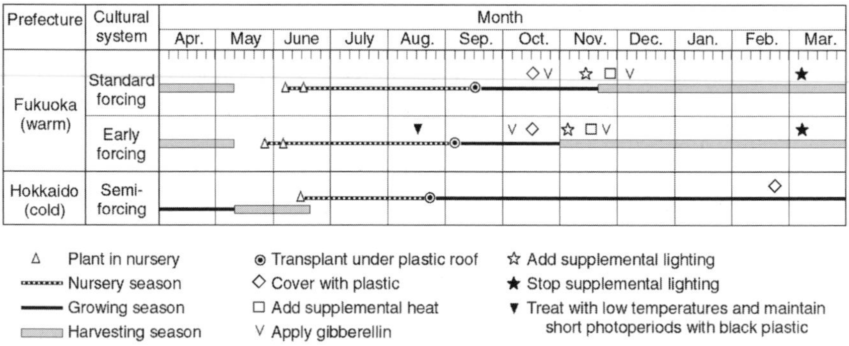

Prefecture	Cultural system	Month											
		Apr.	May	June	July	Aug.	Sep.	Oct.	Nov.	Dec.	Jan.	Feb.	Mar.
Fukuoka (warm)	Standard forcing			Δ···Δ········	·········	·········	⊙	◇V	☆ □ V				★
	Early forcing			Δ·Δ····	·········	▼ ··········	⊙	V ◇	☆ □V				★
Hokkaido (cold)	Semi-forcing			Δ··········	·········	⊙						◇	

Δ   Plant in nursery   ⊙ Transplant under plastic roof   ☆ Add supplemental lighting
········· Nursery season   ◇ Cover with plastic   ★ Stop supplemental lighting
——— Growing season   □ Add supplemental heat   ▼ Treat with low temperatures and maintain
▨▨▨ Harvesting season   V Apply gibberellin   short photoperiods with black plastic

Fig. 6.6. Summary of several types of Japanese forcing culture (redrawn from information provided by K. Kawagishi).

ticularly the eastern and mid-western USA. The bulk of the fruit destined for the fresh market is hand harvested on a 2–3 day cycle, except in extremely hot weather. Little picking is done in heavy rain, but if forced air cooling is available, most producers do not worry much about moisture on the fruit. Harvesting is commonly done from first light until the middle of the day, although some picking is done later in the day.

Pickers are paid by either volume, weight or time, depending on location and the personal preferences of the producers. For example, in Korea and Japan, most pickers are paid by the hour or day (K. Kawagishi, Japan and Ho-joeng Jeong, Korea, 1999, personal communications). In the UK, pickers are paid by weighed tray (D. Simpson, UK, 1999, personal communication). In North America, pickers are commonly paid by volume (quart or tray of pints), but some producers pay by weight. In California, workers are paid by the hour when yields are low so that they meet minimum state and federal minimum wage levels, but during periods of heavy production, they are paid a combination of a base hourly rate plus a piece rate by tray (T. Sjulin and K. Larson, California, 1999, personal communication).

Most fresh market fruit are harvested when they are three-quarters coloured, as these store longer than fully ripe fruit and still develop adequate colour (see Chapter 7). The fresh market fruit is 'snap-picked' with only a little stalk remaining. Processed fruit are often left in the field for longer periods of time and their caps are removed in the field. Most sorting and packing of fresh market fruit is done directly in the field by the pickers. Processed fruit are often picked in large containers in the field and then are sorted over grading lines at the packing house.

The types of containers used to market fresh fruit vary widely across the world. Probably the most widespread container used for commercial markets is the clear plastic clam shell, although plastic mesh and clear plastic containers with non-hinged lids are still popular. Retailers often repack fruit shipped in mesh containers, however, the labour cost of repackaging is becoming prohibitive. Small producers in North America prefer pulp and wooden containers, to signify that the fruit is locally produced.

Container sizes also vary widely. Quarts are commonly used by small producers in North America, whereas larger producers most frequently use 1 lb, containers packed eight to a cardboard tray, or pints packed 12 to a tray. In Europe, fruit are packed in metric containers of 125, 250 and 375 g; 250 is probably the most popular. In Japan, 300 g packages, four to a cardboard box are common.

The number of pickers needed to adequately harvest a field varies widely depending on crop load and cultural system. Experienced pickers can harvest three to four flats of 8 quarts per hour on matted rows, making 10 to 25 pickers necessary per hectare. In annual hill fields, a good picker can harvest 6–10 trays of pints per hour, and about five pickers per hectare are generally needed. In Hokkaido, Japan, three to four pickers are needed per 300 m^2 tunnel.

After harvest, the containers of fresh fruit are palletized in a barn and then commonly forced-air cooled (see Chapter 7). Forced-air cooling involves the channelling of refrigerated air through pallets of fruit (Mitchell, 1994a, b). A pressure gradient is developed across the two sides of a stack of fruit by a fan pulling cold air (as low as $-3°C$) from the room through the spaces in and around the boxes. To minimize fruit dehydration, the fan is stopped when the flesh temperature is within a few degrees centigrade of the target and the cooler is maintained at above 90% relative humidity.

After cooling, fruit are generally kept in a holding room at $2°C$ and high humidity until shipping. Polyethylene wraps are frequently used around pallets of fruit to prevent water loss; however, these can interfere with heat transfer and therefore are not applied until the fruit are fully chilled. To prevent condensation, the pallets are wrapped before they are taken out of cold storage.

Mechanical harvesting

Although virtually all strawberry fruit is harvested by hand, a number of different mechanical harvesters have been developed in western Europe and North America (Dale and Hergert, 1991; Dale et al., 1994). These machines are thought to have the potential to pick fresh quality fruit but, to date, have been used solely for processed fruit (Morris and Cawthon, 1980; Sistrunk and Morris, 1980; Dale and Hergert, 1991). The harvesters are generally used in a field without any hand picking. They have been used experimentally to clean-up fruit after one or two hand harvests (Morris et al., 1979b), but this practice is not economical (Kim et al., 1980; Lauro, 1986).

The machines either strip off fruit or cut them off at ground level (Fig. 6.7). Leaves are separated from the fruit using high speed air currents (Ruff and Holmes, 1976; Morris, et al., 1978). A comb–brush system is utilized to help pick up fruit from the ground and convey it to containers. High density, matted row systems of training are generally used with the machines (Morris et al., 1985a) and fields are sometimes solid set without rows to maximize productivity (Ricketson, 1968; Dale et al., 1987). Fields must be very level to ensure that all the fruit are picked off the ground.

The harvested fruit are then sorted in a processing plant. Several in-plant lines have been developed for purée and whole fruit (Hansen and Ledebuhr, 1980; Kirk, 1980a). Decappers have been developed that either cut or pull calyces from the fruit (Kirk, 1980b; Dale et al., 1987). Continuous flow grading cleaning and grading lines consist of a dump-wash tank to eliminate dirt and rocks, trash eliminator cleaners consisting of rollers revolving in different directions, vibrator-washers for a final wash and tapered finger sizers (Morris et al., 1978). Fruit harvested and cleaned in this manner can be successfully used for juice and purée (Morris, 1980; Sistrunk and Morris,

Fig. 6.7. Schematic drawing of a mechanical harvestor (redrawn from Morris, 1980). a, mowing sickle bar; b, comb-brush picking and conveying system; c, fan; d, air-lock valve; e, fruit transporting conveyor.

1980), and has been successfully frozen and processed into jam with only limited reduction in quality (Morris *et al.*, 1979a, 1980). Postharvest cooling has been shown to improve recovery of quality fruit (Smith, 1986). The primary problem with these products is associated with the high percentage of white and green fruit left after mechanical sorting, but jam can be made from up to 50% green fruit if the anthocyanin content of the other fruit is high.

Although mechanical harvesting systems have shown high promise in many economic studies (Kim *et al.*, 1980; Welch *et al.*, 1986), they have not been widely utilized by growers because it is difficult to consistently recover profitable yields from existing cultivars (Hergert and Dale, 1989). Total yields are reduced by all the fruit being stripped off, regardless of ripeness, and lower percentages of no. 1 grade fruit are recovered than by hand harvesting (50% vs. 90%) (Kim *et al.*, 1980; Dale and Hergert, 1991). There are also problems associated with the efficiency of calyx removal in the processing plant. The non-selectivity of the harvest makes the timing of picking critical to obtaining high proportions of firm, properly coloured fruit (Sistrunk and Morris, 1980). Widespread commercial exploitation of mechanical harvest systems still awaits the development of broadly adapted cultivars with high once-over yields, easy calyx removal, long, strong peduncles and pedicles, and firm, bruise-free fruit with rot resistance.

METHODS OF PROPAGATION

Although strawberries readily proliferate asexually via runners, meristem culture and heat treatment are commonly used for virus elimination before mass propagation (Posnette, 1953; Frazier *et al.*, 1965; McGrew, 1980). Boxus (1974) originally developed the protocol by which millions of plants could be produced from a single meristem. In the first stage, 0.1–0.5 mm meristem domes are excised from newly formed runner tips and are surface sterilized. Long days or gibberellin treatment are used to stimulate runnering (Broome and Goff, 1987). These meristems are generally placed on either Murashige and Skoog (1962) or Boxus (Boxus *et al.*, 1977) basal media. After establishment, the explants are transferred to medium containing a cytokinin to promote axillary bud development and proliferation. After several subcultures, the plantlets are placed into a medium containing auxin to promote rooting or are placed directly into an artificial soil mix. A gradual acclimation period then follows, where plants are moved from the high-humidity *in vitro* environment to greenhouse conditions.

Stock plants are generally heat pretreated at 36°C for about 6 weeks for virus elimination (Posnette, 1953; Converse, 1979). Individual plantlets from each cultured meristem are tested to verify that viruses have been removed. Classically this was done by grafting to indicator stocks sensitive to specific pathogens (Bringhurst and Voth, 1956; Frazier, 1974), but more recently, ELISA (enzyme-linked immunosorbent assay) tests and DNA hybridizations are being used (Stenger *et al.*, 1988; Kaden-Kreuzinger *et al.*, 1995; Martin, 1995; Schoen and Leone, 1995).

Occasional concerns have been raised about genetic stability and the proliferation of mutants (Scaeffer *et al.*, 1980; Sansavini *et al.*, 1990). Permanent genetic changes have been noted in the form of somaclonal variation, with changes in both morphological traits and pathogen resistance being observed (Swartz *et al.*, 1981; Boxus *et al.*, 1984; Shoemaker *et al.*, 1985). High numbers of repeated subcultures appear to amplify this possibility (Kinet and Parmentier, 1989), and as a result, a limit of five to ten subcultures has been recommended.

Epigenetic differences that disappear over time have also been noted between *in vitro* propagated plants and their conventionally propagated counterparts, presumably due to the hormones contained in the culture media. Tissue cultured plants often produce more runners (Boxus *et al.*, 1984), and in some instances, have higher photosynthetic rates (Swartz *et al.*, 1981; Cameron and Hancock, 1985, 1986). These changes can lead to higher yields, if plant densities do not increase to the point of crowding, although fruit size is often negatively impacted. To minimize these effects, multiple daughter generations have proven helpful, with a minimum of two field propagations.

Programmes for the production of certified virus free, true-to-type plants have been developed in many countries (Goff, 1986). In general, several steps are employed (Broome and Goff, 1987; Fiola, 1996). First, 'nuclear stock' representing virus tested plants out of tissue culture are planted in a greenhouse or screenhouse where insects are excluded. The first-year daughters produced by these plants (registered or foundation stock) are set in a fumigated field to produce another round of daughter plants for sale to commercial nurseries. The daughter plants produced from these plants are sold to growers as 'certified stock'.

As was previously mentioned, planting stock is dug as actively growing green plants, or as dormant or semidormant plants whose full dormancy requirement has not been met. Green plants are dug directly from production fields or by rooting runners (plug plants) under mist in the greenhouse or field (Poling and Parker, 1990; Poling, 1994; Bish *et al.*, 1997). Fully dormant plants are generally dug in early winter and stored until planting (Anderson, 1982). Plants with variant levels of chilling are generated by digging plants from the field at different times in the autumn and at different geographical locations (Faby, 1997; Chandler *et al.*, 1989; Bringhurst and Voth, 1990). For example, both high elevation and high latitude nurseries are utilized in California to provide plants for planting in the south.

Dormant plants are generally stored bare-rooted in unsealed, polyethylene lined boxes or crates at slightly below freezing temperatures (Worthington, 1970; Anderson, 1982). Sphagnum moss or similar packing materials are not necessary if polyethylene liners are used (Bryant *et al.*, 1961). The optimum temperature for long-term storage is $-1°C$, as temperatures colder than this can cause injury to crown tissues (Guttridge *et al.*, 1965; Anderson, 1982; Lieten and Goffings, 1997). Controlled atmospheres have been shown to increase the vigour of stored plants and reduce decay (Lockhart and Eaves, 1966; Lockhart, 1967), although reduced-O_2 conditions have not always proven beneficial (Lieten and Goffings, 1997).

CONCLUSIONS

There are two major strawberry cultural systems – perennial matted rows and annual hills. Matted rows exploit runners as the primary yield component and are most commonly used in climates with short summers and cold winters. Hill culture relies on crowns as the primary yield component and are used predominantly in areas having warm winters and either hot or moderate summers. The annual hills feature summer or autumn/winter planting, raised beds, plastic mulch, trickle irrigation and 1–2 year production seasons. The matted row systems feature spring planting, flat beds, straw mulch, overhead irrigation and three to five production seasons. The annual hills are much more productive than the matted rows, but the reduced number of inputs utilized in matted rows allows them to be profitable in climates where hills can not be grown successfully.

There is considerable interest in growing annual hills as far north as possible to exploit their higher yields and better weed control. In the mid-Atlantic coast of the US and the UK, strawberries are now grown in a number of areas where matted rows were previously utilized exclusively. The further expansion into northern regions will depend on the availability of cold adapted, annual cultivars and their ability to produce economic numbers of flower buds in the shorter northern summers.

Although matted row culture is fairly standard across the world, annual hill culture varies dramatically depending on the environmental and economic realities in different areas. The earliest, highest quality fruit are produced in late autumn/early winter planting systems, whereas the highest yields and dependability come from summer planting systems. Clear plastic mulches are used in areas where winter temperatures are a little too cool for sustained growth and black plastic is used in areas where winter temperatures are adequate but summer temperatures can be excessive.

Season extension has come through the widespread use of day-neutral types in the warm regions of the USA, and through tunnels and greenhouses in Europe. It is expected that the popularity of day-neutrals will continue to grow and reduce the use of protected culture in Europe; however, tunnels will probably retain a key role in expanding the season of high quality local varieties in both continents.

Weed control has been a long-standing problem for matted row growers and could also become a problem for annual hill growers when the use of methyl bromide is banned in some of the major producing regions. It is likely that chemical alternatives to methyl bromide will be found, but non-chemical techniques of weed control will also become increasingly important. Numerous techniques show promise including the use of living mulches, planting into killed sod and preplant cover crops. Transgenic crops with herbicide resistance will also become important.

7

FRUITING AND POSTHARVEST PHYSIOLOGY

INTRODUCTION

Ripe strawberries are composed of approximately 90% water and 10% total soluble solids (Hemphill and Martin, 1992), and contain numerous important dietary components (Table 7.1). They are extremely high in vitamin C and a standard serving of strawberries (ten fruits) supplies 95% of the recommended dietary requirements (National Academy of Sciences, 1989; Maas *et al.*, 1996). The main soluble sugar components in strawberries are glucose and fructose, which comprise over 80% of the total sugars and 40% of the total dry weight (Wrolstad and Shallenberger, 1981). The primary organic

Table 7.1. Composition (per 100 g fresh weight) of fresh, unprocessed strawberry fruit (Maas *et al.*, 1996).

Nutrient	Content	Nutrient	Content
Water	92 g	Vitamins	
Food energy	30 kcal	Vitamin C	56.7 mg
Protein	0.6 g	Others	<0.5 mg
Lipids, total	0.4 g		
Carbohydrate, total	7.0 g	Lipids	
Fibre	0.5 g	Saturated	0.020 mg
Ash	0.4 g	Monounsaturated	0.052 mg
Minerals (mg)		Polyunsaturated	0.186 mg
Ca	14 mg	Cholesterol	0
Fe	0.4 mg	Phytosterols	12 mg
Mg	10 mg		
P	19 mg	Amino acids	
K	166 mg	All (*n* = 18)	522 mg
Na	1 mg		
Zn, Cu, Mn	<0.5 mg		

acid is citric acid, which comprises 88% of the total acids (Green, 1971). The strawberry also contains significant levels of ellagic acid, which is thought to be an anticarcinogenic (Maas *et al.*,1991a, b, 1996).

Strawberry flavour is a complex combination of sweetness, acidity and aroma. The most intensely flavoured fruit generally have high levels of both titratable acidity (TA) and soluble solids, whereas the most bland fruit are low in both these components (Kader, 1991). The primary components of flavour have not been completely elucidated, but strawberry aroma is thought to originate from a complex mixture of esters, alcohols, aldehydes and sulphur compounds (Schreier, 1980; Dirinck *et al.*, 1981; Pérez *et al.*, 1996).

Since strawberries are such a widely grown commodity, much research has been conducted on producing and maintaining these standards of fruit quality. Considerable effort has been expended on maximizing fruit growth in the field and increasing postharvest fruit quality. Almost 40% of the fruit can be lost after picking, mostly to decay (Ceponis *et al.*, 1987), so postharvest handling is critical. In many cases, fresh strawberries are shipped long distances by air or truck, making the length of shelf life a critical parameter. Maintaining maximum fruit quality in processed products is also an area of intense interest. This chapter discusses the various factors influencing fruit growth, and addresses the biochemistry of ripening and postharvest handling procedures.

FACTORS INFLUENCING FRUIT GROWTH

Strawberry fruit growth as a function of time fits a single sigmoid curve or is biphasic depending on cultivar (Mudge *et al.*, 1981; Stutte and Darnell, 1987; Miura *et al.*, 1994). Cells in the cortex and pith are responsible for most of the receptacle growth, with the cortex being the primary contributor to fruit size. Cell division accounts for only 15–20% of the total growth, occurring mostly before anthesis. The rest of the growth is a result of cell enlargement, with cell size increasing towards the inner part of the fruit. Sugars, aromatic compounds and pigments all increase as the receptacle tissue matures. Ripening from anthesis to harvest averages about 30 days depending on environmental conditions.

Fruit development is affected by numerous factors including the number of achenes on the receptacle (Moore *et al.*, 1970), the area of receptacle tissue surrounding each achene and the distribution of the achenes on the receptacle. To produce a well-shaped berry, it is necessary that a minimum of 30% of the carpels are fertilized (Day, 1993). Malformed fruit are caused by several factors including poor pollination and damage to the achenes by frost, insect or disease. All of these prevent the synthesis of auxin and result in uneven development. Cool temperatures can also have a negative impact on fruit development by reducing pollinator activity and limiting pollen production (Risser, 1997).

Honey bees (*Apis mellifera*) are the most effective pollinators of strawberries in the open field (Antonelli *et al.*, 1988; Goodman and Oldroyd, 1988; Pritts and Handley, 1998), although the activity of other bees can be complementary (Chagnon *et al.*, 1993). The bumble bee (*Bombus terrestris*) is commonly employed as a pollinator in protected cropping. Numerous sugars and free amino acids exist in the nectar and pollen of strawberry flowers, but none were directly associated with attractiveness to honey bees (Grunfeld *et al.*, 1989).

Growth of the receptacle is controlled primarily by auxin, which is synthesized in the achenes (Nitsch, 1950; Dreher and Poovaiah, 1982; Archbold and Dennis, 1984). Fruit development in some cultivars responds to a wide range of compounds with auxin-like activity (Mudge *et al.*, 1981), whereas in several day-neutral types, only the ethyl-ester of indolyl acetic acid (IAA) is active (Darnell and Martin, 1987). The partial removal of achenes at early stages of development results in distorted berry formation, with the receptacles expanding only in proximity to undisturbed achenes (Fig. 7.1).

Fig. 7.1. Growth of strawberry fruit when achenes have been removed: (A) one single fertilized achene, (B) three fertilized achenes, (C) three rows of fertile achenes, and (D) all achenes fertilized.

After total seed removal, the application of β-naphthylacetic acid allows fruit expansion to continue when applied at 4, 7, 12, 19 and 21 days post-pollination. Auxin has been found to stimulate the appearance of two mRNAs, SAR1 and SAR2, that are correlated with fruit expansion in pollinated and unpollinated fruit (Reddy *et al.*, 1990).

Although auxin stimulates the swelling of receptacles, it inhibits the overall ripening process. Detachment of achenes from green fruit hastens ripening, as measured by anthocyanin accumulation (Given *et al.*, 1988a; Manning, 1997). When the synthetic auxin 1-naphthylacetic acid (1-NAA) is applied to detached fruit, the development of red coloration is slowed compared to water controls and fruit having applications of the inactive auxin analogue phenoxyacetic acid (POA). An important factor in triggering ripening is probably the natural decline in auxin content that occurs during achene maturation (Archbold and Dennis, 1984).

Strawberry fruit are the most competitive sink in the plant, accumulating 20–40% of the total plant dry weight (Forney and Breen, 1985; Schaffer *et al.*, 1986). Fruiting limits dry matter accumulation in roots, crowns and leaves, and inhibits runner, crown and inflorescence production. However, fruiting generally does not affect total levels of dry weight accumulation in the plants (Forney and Breen, 1985; Schaffer *et al.*, 1986b) and the only tissues to show an actual reduction in biomass during fruiting are the roots (May and Pritts, 1994).

During rapid periods of fruit growth, fruit dry weight accumulation may exceed the assimilatory capacity of the plant, and continued fruit growth is maintained by translocation from other plant parts (Antoszewski and Dzieciol, 1973; Jurik, 1983). Darnell and Martin (1988) estimated that only 25% of the carbon required for the first 7 days of fruit development is supplied by current photoassimilates. Translocation of ^{14}C-incorporated photosynthates to fruit increases as fruit development progresses, with primary flowers being a stronger sink than tertiary flowers (Nishizawa and Hori, 1988). Sucrose uptake is greatest among young fruit (2 days after pollination) compared with older fruit (8–27 days after pollination) (Archbold, 1988). Soluble sugars and starch levels increase in leaves between fruit harvest and late autumn, then decline as they are mobilized to crowns and roots, and then are mobilized into developing flowers and fruits (Long, 1935).

The carbon cost for fruit is so high that net CO_2 balances are negative (Jurik, 1983). Forney and Breen (1985) found decreases in petiole and root dry weight during early fruit development in short-day plants, and Schaffer *et al.* (1986b) discovered that dry weight accumulation in the leaves of fruiting day-neutrals was reduced. In forcing culture, it has been shown that the developing fruit of short-day plants accumulate 31% of the total dry plant weight (Nishizawa, 1994). Bearing plants make less vegetative growth than non-bearing plants and their total dry weight accumulation is smaller after 124 days. Pritts and Worden (1988) discovered that early fruit removal enhances the yields of day-neutral types, and Hancock and Cameron (1986) found that harvesting in the first year has a negative impact on subsequent yields in a number of short-day cultivars.

Elevated temperatures have a negative effect on fruit size and quality, as strawberries have a high respiration rate, high surface to volume ratio and a thin cuticle (Perkins-Veazie, 1996). Fruit temperatures can exceed air temperatures by as much as 8°C on sunny days of 26.5°C (Austin *et al.*, 1960), causing tissue damage, softness and breakdown near the berry surface. Temperatures above 25°C can also reduce fruit set (Kronenberg *et al.*, 1959; Abdelrahman, 1984), decrease fruit soluble solids content (Abdelrahman, 1984; Hellman and Travis, 1988) and increase the rate of fruit development (Darrow, 1966; Dana, 1980). Went (1957) showed that daytime air temperatures above 15–17°C reduced fruit size and fruit aroma, and optimal fruit growth occurred at 12°C night temperatures. Miura *et al.* (1994) found that rates of fruit growth were faster at 19 than 15°C air temperatures, but total fresh weight and dry weight per inflorescence was much lower at 19°C. Fruit size in several day-neutral types in Maryland was about 50% smaller in the hot summer than cool spring (Draper *et al.*, 1981), and Galletta *et al.* (1981b) suggested that a decrease in soil temperature of 10°C can increase fruit size by 0.9–1.6 g per berry.

FRUIT RIPENING PROCESSES

Numerous biochemical changes occur during fruit ripening (Fig. 7.2; Manning, 1993). Over 50 polypeptides have been identified that show prominent changes at different stages of fruit development (Manning, 1994; Reddy and Poovaiah, 1987; Veluthambi and Poovaiah, 1984). Several specific enzymes associated with membranes (Civello *et al.*, 1995), anthocyanin synthesis (Given *et al.*, 1988b, c) and sucrose metabolism (Hubbard *et al.*, 1991) have been shown to increase in the strawberry during ripening.

The mRNA composition changes during ripening from immature to green fruit, with the greatest changes occurring just before ripening (Manning, 1994, 1997). Total RNA from receptacles of immature green fruit has been used to clone for two auxin-induced mRNAs (Reddy *et al.*, 1990) and one auxin-repressed mRNA (Reddy and Poovaiah, 1990). From a cDNA library prepared from mRNA isolated from ripe fruit, a number of homologies have been identified including several gene families putatively encoding enzymes of phenylpropanoid metabolism, and genes for cellulase, expansins, cysteine proteinase and acyl carrier protein (Manning, 1998; Table 7.2). Three mRNAs with fruit specific, ripening-enhanced expression have also been identified in ripening fruit using polymerase chain reaction (PCR) differential display. When sequenced, they had high homology with known proteins including: (i) an annexin which may play a role in membrane function and cell wall structure, (ii) chalcone synthase which is a key enzymatic step in flavonoid

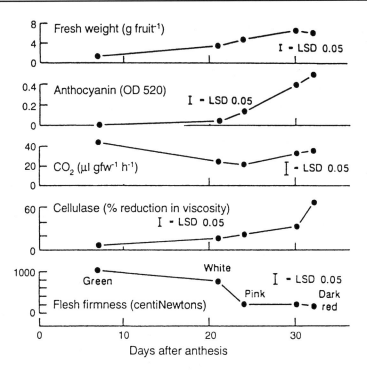

Fig. 7.2. Changes in fruit weight, anthocyanin content, CO_2 production, cellulase activity and flesh firmness during strawberry fruit ripening (redrawn from Abeles and Takeda, 1990).

Table 7.2. Ripening related genes that have been cloned by Manning (1997, 1998).

Gene	No. of families	Metabolic pathway	Quality attribute
O-methyl transferase	1	Phenylpropanoid	Colour
Chalcone synthase	2	Phenylpropanoid	Colour
Flavanone-3-hydrolase	4	Phenylpropanoid	Colour
UDP-glucosyl transferase	1	Phenylpropanoid	Colour
UDP-glucuronosyl transferase	1	Phenylpropanoid	Colour
Cellulase	1	Cell wall hydrolysis	Texture
Acyl carrier protein	2	Fatty acid biosynthesis	Flavour
Cysteine proteinase	1	Protein degradation	Ripening

biosynthesis, and (iii) a ribosomal protein, most likely a 40S subunit (Wilkinson *et al.*, 1995). In addition, a gene (*njjs*4) has been identified which is associated with the process of seed maturation and fruit ripening, and is related to the class-I low-molecular-weight heat-shock-protein-like genes (Medina-Escobar *et al.*, 1998).

Strawberries are not climacteric, as they produce little ethylene (Knee *et al.*, 1977; Abeles and Takeda, 1990), and as a result, the application of ethylene has little effect on the softening and flavour development of immature fruit (Abeles and Takeda, 1990; Perkins-Veazie, 1995). In addition, the onset of ripening measured by anthocyanin accumulation is not slowed by inhibitors of ethylene synthesis (aminoethoxyvinylglycine) or of ethylene action (norbornadiene, silver) (Given *et al.*, 1988a). There is one report that white–pink fruit exposed to ethylene do have increased levels of total RNA, poly(A)$^+$ RNA levels and RNase activity, suggesting that although ethylene does not initiate a climacteric response, it may still be involved in levels of gene expression promoting some characteristics associated with ripening (Lou and Liu, 1991).

Strawberry softening is highly correlated with cultivar and preharvest environment (Kader, 1991; Ponappa *et al.*, 1993), and is strongly associated with the release of pectins and hemicelluloses (Knee *et al.*, 1977; Huber, 1984). Expansins, a class of protein associated with cell expansion, are also thought to play a role in firmness (Manning, 1998). The strawberry has little endo-polygalacturonase, the most important enzyme in tomato ripening (Nogata *et al.*, 1993); instead the most important softening enzymes are thought to be pectinmethylesterase and cellulase (Huber, 1984; Abeles and Takeda, 1990).

Red colour develops through the production of anthocyanins, primarily pelargonidin-3-glucosidase (Pg 3-gl) (Timberlake and Bridle, 1982). Almost 90% of the anthocyanins are Pg 3-gl (Wrolstad *et al.*, 1970; Kalt *et al.*, 1993), although at least eight pelargonidin- and two cyanidin-based anthocyanins have been detected in strawberry juice (Bakker *et al.*, 1994). Cyanidin 3-glucoside (Cy 3-gl) is the second most common anthocyanin. The total concentration of anthocyanins varies 16-fold across varieties, and there is some variation in anthocyanin composition, although no clear associations between individual anthocyanins and colour have been observed (Bakker *et al.*, 1994). Overripe, purplish fruit result when cell compartmentation is lost and there is a shift in anthocyanin glycosylation due to acidic to basic pH changes (Manning, 1993).

Anthocyanins start to appear during the white stage of fruit development (Fig. 7.3), when mRNAs from several genes associated with phenylpropanoid metabolism are up-regulated (Manning 1994, 1998). A key step is the *de novo* synthesis of phenylalanine-ammonia lyase (PAL), paralleled by the appearance of the terminal enzyme in the synthesis of Pg-3-gl, uridine diphosphate glucose:flavonoid O^3-transferase (UDPGFT) (Given *et al.*, 1988b, c). Activity of PAL has been shown to have two peaks during fruit ripening: (i) when there are maximum levels of soluble phenols at the green fruit stage 5 days after anthesis; and (ii) when the fruit are nearly ripe at 27 days after anthesis

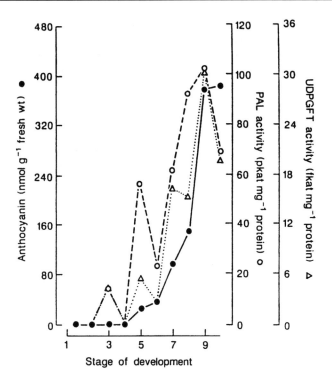

Fig. 7.3. Anthocyanin content, phenylalanine ammonia-lyase (PAL) and uridine diphosphate glucose: flavonoid O^3-transferase activities in ripening fruit (redrawn from Given *et al.*, 1988a).

(Cheng and Breen, 1991). Other important enzymes associated with anthocyanin production are O-methyltransferase, chalcone synthase, chalcone reductase, flavanoid-3-hydroxylase, UDP-glucosyl transferase and UDP-glucuronosyl transferase (Manning, 1998).

Soluble solid content continually increases during strawberry development, from 5% in small green fruit to 6–9% in red berries (Spayd and Morris, 1981; Kader, 1991). Green and red fruit vary little in their pH, but titratable acidity gradually declines during ripening. Like all other fruit quality parameters, soluble solids and titratable acidity are dependent on cultivar and environmental conditions.

Hundreds of volatile esters have been identified during strawberry ripening and aroma development (Table 7.3; Zabetakis and Holden, 1977; Pérez *et al.*, 1992), with methyl and ethyl esters of butanoic and hexanoic acids being among the most prevalent (Larsen and Poll, 1992; Pérez *et al.*, 1992, 1993, 1996). Other components in high concentration are *trans*-2-hexenyl acetate, *trans*-2-hexenal, *trans*-2-hexenol and 2,5-dimethyl-4-hydroxy-3(2H)-furanone (furaneol) (Schreier, 1980). Concentrations of these volatiles vary widely

Table 7.3. Some representative compounds for primary odour characteristics (POC) of strawberries (Scheerens and Stetson, 1996).

POC	Chemical family	Representative compounds
Fruity	Low mol. wt esters	Ethyl butanoate
		Ethyl hexanoate
		Hexyl acetate
		Isoamyl acetate
	Ketones	2-Heptanone
	Lactones	γ-Decanolactone
Floral/citrus	Terpenols	Linalool
		α-Terpineol
Burnt sugar	Furanones	Furaneol
	Ketones	3-Hydroxybutanone
Spicy	Various	Ethyl cinnamate
Buttery	Low mol. wt diones	Diacetyl
Nutty	Benzene derivative	Benzaldehyde
Herbaceous	Unsat. aldehydes	*t*-2-Hexenal
	Unsat. alcohols	*t*-2-Hexen-1-ol
	Unsat. esters	*t*-2-Hexenyl acetate
Baked/cooked	Furaldehydes	Furfural
Fatty/waxy/coconut	Lactones	γ-Caprolactone
	Inter. wt alcohols	Lauryl alcohol
Rancid	Sat. aldehydes	Hexanal
Goaty	Vol. fatty acids	2-Methylthiol acetate
Sulphurous	Thiol esters	Methylthiol acetate
Chemical	Various	Naphthalene

among cultivars (Hirvi, 1983; Shamaila *et al.*, 1992; Pérez *et al.*, 1996, 1997) and produce large variations in aroma quality. 'Tudla' and 'Oso Grande' were found to have the highest aroma quality among a number of cultivars grown in Spain, whereas 'Seascape' and 'Cuesta' had the lowest (Pérez *et al.*, 1997).

Aroma and fragrance content also varies across species (Hirvi and Honkanen, 1982; Mochizuki *et al.*, 1997). The wild species *F. vesca* and *F. virginiana* have much stronger aroma than the cultivated types (Hirvi and Honkanen, 1982). *F. vesca* contains high amounts of ethyl acetate, but low amounts of methyl butyrate, ethyl butyrate, and furanone. *Fragaria nilgerrensis* contains high levels of ethyl acetate and furanone, but low levels of methyl butyrate and ethyl butyrate. Hybrids between *F. vesca* and *F.* × *ananassa* have intermediate levels of fragrance and aroma, and crosses between *F. nilgerrensis* and *F.* × *ananassa* more closely resemble *F. nilgerrensis*.

Several researchers consider 2,5-dimethyl-4-hydroxy-3(2H) furanone (furaneol) and 2,5-dimethyl-4-methoxy-3(2H) furanone (mesifurane) as being particularly important aroma contributors (Pyysalo *et al.*, 1979; Larson and Poll, 1992; Sanz *et al.*, 1994; Pérez *et al.*, 1996). Larson and Poll (1992) have demonstrated that the combination of furaneol and ethyl butanoate produce a strawberry-like odour. During natural ripening of fruit, the furaneol, mesifurane and furaneol glycoside content increases with the highest concentrations in overripe fruit (Pérez *et al.*, 1996). Furaneol is the only furanone detected in white fruit, and it dominates in all ripening stages. During storage in a modified atmosphere, concentrations of furaneol drop while mesifurane and furaneol glycosides increase.

The identification of enzyme pathways in aroma and flavour production are beginning to emerge. Pérez *et al.* (1993, 1996) has identified an alcohol acyltransferase as one of the critical enzymes in volatile synthesis. Mitchell and Jelenkovic (1995) have suggested that NAD- and NADP-dependent alcohol dehydrogenase activity are important in flavour and fragrance activity in ripening strawberries. In Northern analyses, Manning (1998) has found levels of the acyl carrier protein (ACP) to rise markedly at the onset of ripening. He suggests that in the aroma pathways, short-chain fatty acids are converted into aldehydes and alcohols by the alcohol dehydrogenases, and esters are formed from these alcohols by the acyl transferases.

Glucose, fructose and sucrose are the major soluble sugars found in the fruit of strawberries during all stages of ripening. Glucose and fructose are found in almost equal concentrations (Maas *et al.*, 1996), and they rise continuously during fruit development from 5% in small green fruit to 6–9% in red berries (Kader, 1991; Spayd and Morris, 1981). Sucrose levels are generally much lower, and show little accumulation until about the middle of fruit development (Forney and Breen, 1986). Invertases probably play an important role in regulating sweetness, by regulating hexose and sucrose levels (Ranwala *et al.*, 1992; Manning, 1998).

The pH of strawberry fruit remains at about 3.5 during fruit development (Spayd and Morris, 1981), although titratable acidity, representing predominantly organic acids, gradually drops during fruit development. Levels of both sugars and acids vary greatly in ripe fruit, depending on cultivars and developmental conditions (Table 7.4).

HANDLING AND STORAGE OF FRESH FRUIT

The highest quality fresh fruit are regularly shaped, glossy, fully coloured, firm and have a healthy green calyx. They are light-red to orange–red, juicy, aromatic and have no mould or bruises. Size is also important, but fruit that are too large can be damaged by package covers.

The nutrient status of plants in the field can strongly influence fruit quality. Too much nitrogen can result in smaller fruit, more fruit decay, higher fruit respiration, and decreased soluble solid content (SSC), flavour and firmness (Overholser and Claypool, 1931; Kader, 1991; Perkins-Veazie and

Table 7.4. Levels of sugars and non-volatile carboxylic acids typically found in ripe strawberry fruits (adapted from Maas *et al.*, 1996).

Tastant	Reported ranges (g per 100 g fresh fruit)
Sugars	
Fructose	1.0–3.5
Glucose	1.4–3.4
Sucrose	0.2–2.5
Amino acids	
Aspartic	Trace – 0.03
Glutamic	Trace – 0.04
Carboxylic acids	
Citric	0.29–1.24
Malic	0.09–0.68

Collins, 1995). Calcium has long been implicated in fruit firmness and rot resistance, but applications have had variable results depending on cultivar, rate and timing (Cheour *et al.*, 1990, 1991; Chung *et al.*, 1993). Foliar applications 3–4 days before harvest of 20 kg ha^{-1} calcium chloride, reduced fruit softness and decay during storage and fruit stored 7 days at 5°C had less anthocyanin and slightly higher soluble solids (Cheour *et al.*, 1990). Lieten and Marcelle (1993) have associated soft, albino fruit with excessive nitrogen, low calcium and potassium and insufficient light.

Fruit must be picked when they are at least 25% ripe or pink for acceptable development of colour and volatile production in storage (Smith and Heinze, 1958; Miszczak *et al.*, 1995). In general, it is recommended that fruit be picked at least three-quarters ripe, as fruit stored at this stage develop adequate colour and are firmer after storage than those harvested fully ripe (Pritts *et al.*, 1987; Mokkila *et al.*, 1997). Strawberries should not be picked overripe, as colour, both externally and internally, darkens during storage and becomes more intense (Sacks and Shaw, 1993). Glossiness also diminishes in storage, particularly at temperatures above 5°C and low humidity (Collins and Perkins-Veazie, 1993; Ferreira *et al.*, 1994).

The anthocyanin composition of stored fruit differs from field ripened fruit, with there being a higher proportion of cyanidin 3-glucoside (Cy 3-gl) in storage (Kalt *et al.*, 1993). This may be why stored fruit are darker, as Cy 3-gl is much darker magenta than the orange–red of Pg 3-gl (Harborne, 1984; Kalt *et al.*, 1993). Although light plays an important role in anthocyanin production in the field, it has at best, a modest influence on colour development in storage (Kalt *et al.*, 1993; Saks *et al.*, 1996).

Once picked, the removal of field heat as soon as possible is critical to preserving fruit quality. Rapidly cooled fruit can maintain acceptable quality in storage for 5–10 days, if held below 5°C at high relative humidity (>90%) to minimize fruit weight loss (Mitchell *et al.*, 1964). For every 3 h delay in cooling to 5°C, decay has been found to double (Maxie *et al.*, 1959). Fruit held at 30°C for 6 h before cooling, loses 4% more weight and is 20% softer than fruit chilled immediately (Nunes *et al.*, 1995). Strawberries are most commonly cooled either passively by placement in a refrigerated room or with forced-air (pressure cooling). Forced-air is the method of choice, as it cools the fruit much more rapidly (see Chapter 6 for details). Hydrocooling fruit by immersion in chlorinated water has also been shown to be a promising method as it reduces grey mould decay and causes less fruit weight loss than forced-air cooling (Ferreira *et al.*, 1996), but has not been accepted commercially.

Since high CO_2 concentrations (15–20%) reduce fungal growth and help maintain quality, strawberries are frequently held in controlled atmosphere storage to increase shelf life. Typical gains in storage life compared to normal air are 3 to 5 days. Controlled atmosphere at elevated CO_2 (10–20%) and decreased O_2 (5–10%) can also lower decay and increase firmness (El Kazzaz *et al.*, 1983; Li and Kader, 1989; Ke *et al.*, 1991; Smith, 1992). Responses to controlled atmosphere storage vary across cultivar and growing region (Brown *et al.*, 1984; Aharoni and Barkai-Golan, 1987; Smith and Skog, 1992).

Modified atmospheres are often maintained after palletization, for long distance travel (Mitchell, 1992). Pallets are wrapped in plastic, sealed, the air is removed and then an atmosphere composed of a mixture of N_2 and CO_2 is injected to obtain CO_2 concentrations of 15–20% (Kader, 1991). CO_2 levels range from 4 to 26% during shipment, with the effectiveness depending primarily on the quality of the seal (Harvey, 1982). CO_2 has also been added to pallets of fruit by placing dry ice in them before wrapping (Winter *et al.*, 1940; Harvey *et al.*, 1971).

Attempts have been made to maintain the quality of fruit by wrapping consumer units with selectively permeable or perforated films (Miller *et al.*, 1983; Aharoni and Barkai-Golan, 1987; Picon *et al.*, 1993; Larsen and Watkins, 1995). Under constant temperature conditions, these modified atmosphere packages (MAP) will slow moisture loss, reduce oxygen concentration and increase CO_2 concentration (Perkins-Veazie, 1995), but under temperature fluctuations anaerobic conditions can develop rapidly and have prevented the commercial use of these films (Kader, 1994).

Off-flavours can occur under both MAP and controlled atmosphere storage conditions with either low O_2, high CO_2 or extended periods of storage (Harris and Harvey, 1973; Ke *et al.*, 1991; Shamaila *et al.*, 1992). The major causes of these off-flavours are the buildup of ethanol, acetaldehyde and ethyl acetate (Fig. 7.4; Li and Kader, 1989; Ke *et al.*, 1991).

Concentrations of many volatiles drop in modified and controlled environments (Pérez *et al.*, 1996), and several different kinds of off-flavours and odours can arise in no more than 12 h, if precise atmospheric conditions

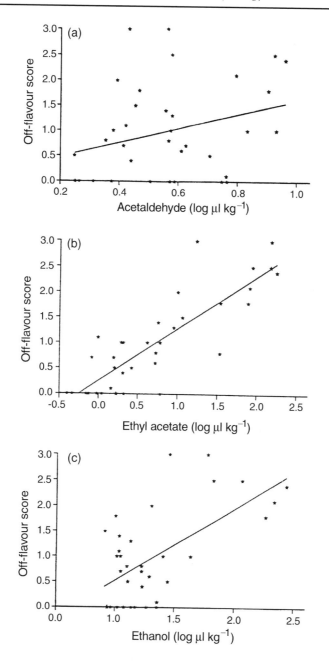

Fig. 7.4. Relationship between off-flavour scores in stored strawberries and content of (a) acetyladehyde, (b) ethyl acetate and (c) ethanol (redrawn from Larsen and Watkins, 1995).

are not maintained. Under stress levels of low O_2 or elevated CO_2, increases in acetaldehyde are common, which degrade rapidly to alcohol (Li and Kader, 1989; Ke *et al.*, 1991; Ueda and Bai, 1993). A number of aromatics with off-tastes can also build up, particularly ethyl acetate (Ueda and Bai, 1993; Larson, 1994; Larsen and Watkins, 1995). Pérez *et al.* (1996) have suggested that levels of alcohol acetyl transferase (AAT) play a key role in the retention of flavour and aroma during storage in CO_2, perhaps due to a detoxifying role.

A number of chemical treatments and microbial antagonists have been identified that can extend the postharvest life of fruit, but these are still in the experimental stage. Pyrrolnitrin from *Penicillium cepacia* Burkh. has been shown to prevent grey mould growth for 5 days on strawberries stored at 1°C (Takeda *et al.*, 1990). Isolates from *Trichoderma viride* Pers. Fr. and *Gliocladium roseum* also inhibit grey mould sporulation and growth (Peng and Sutton, 1991). Chitosan, a semi-permeable polysaccharide, has been shown to reduce decay and slow post-harvest ripening when applied to half-ripe strawberries (El Ghaouth *et al.*, 1991). Fumigation with low concentrations of acetic acid can be used to reduce rot (Moyls *et al.*, 1996). The firmness and percentage decay of whole and sliced strawberries after storage has been enhanced by postharvest dips in 1% calcium chloride (Morris *et al.*, 1985b; Rosen and Kader, 1989; Garcia *et al.*, 1996a). In addition, the polyamines, spermine and spermidine, have been shown to increase the firmness of strawberry slices when vacuum infiltrated, comparable to calcium chloride (Ponappa *et al.*, 1993).

A number of natural volatilization compounds have also been examined experimentally for their antifungal properties. Acetaldehyde gas has been shown to reduce decay (Prasad and Stadelbacker, 1974), but it also decreases acidity and soluble solids content, and can result in off-flavours at high concentrations (Vaughn *et al.*, 1993; Pesis and Avissar, 1990). Vaughn *et al.* (1993) identified five natural volatile compounds that completely inhibited the growth of *Alternaria alternaria, Botrytis cinerea* and *Colletotrichum gloeosporioides*. Benzylaldehyde was effective at the lowest concentrations in the vapour phase (0.04 µl ml^{-1}), but 1-hexanol, E-2-hexenal and 2-nonanone were also effective at 0.1 µl ml^{-1}. Archbold *et al.* (1997a, b) also found several volatile compounds that effectively controlled botrytis rot on strawberry fruit, including benzylaldehyde, methyl benzoate, methyl salicylate, 2-nonanone, E-2-hexenal, diethyl acetal, 1-hexanol and E-2-hexen-1-ol. These compounds vary greatly in their effects on fruit quality and the length of exposure necessary to control rot. Although (E)-2-hexenal has significant antifungal activity at concentrations of 85.6 µM, lower concentrations actually stimulate mycelial growth (Fallik *et al.*, 1998).

Brief heat treatments after harvest have been tested as a means of minimizing postharvest losses. Hot water treatments of 44 or 46°C for 15 min delayed the development of *B. cinerea* in inoculated fruits with minimal loss of quality (Garcia *et al.*, 1996b). 'Tudla' fruit subjected to 45°C for 15 min and then stored at 1°C for 2 days had better overall quality than those treated at 25, 35 or 55°C (Garcia *et al.*, 1995). They had the lowest postharvest losses, weight losses and titratable acidity, together with the highest values for fruit firmness, soluble solids and sensorial appearance. However, the heat treatments diminished calyx colour and fruit skin brightness.

Irradiation has been shown to slow postharvest ripening and reduce decay in strawberries, but has not been utilized commercially. Doses of 1 and 2 kGy can extend strawberry shelf life by several days through the control of fungal rots (Maxie and Adel-Kader, 1966; Yu *et al.*, 1995). At least 1 kGy of irradiation is needed to eliminate *Rhizopus* and 2 kGy to reduce *Botrytis* (Brecht *et al.*, 1992). Unfortunately, doses this high can result in reduced colour (Thomas, 1986; Couture *et al.*, 1990) and fruit softening through changes in the oxalate-soluble pectase fraction (Yu *et al.*, 1996). Since high levels of irradiation can not be used on strawberries without negatively impacting on their fruit quality, irradiation may be most effective if it is combined with other postharvest techniques such as modified atmosphere packaging (Brecht *et al.*, 1992) or heat treatments (Sommer *et al.*, 1968).

PROCESSING FRUIT

Strawberries are utilized in a number of products including preserves, fruit yoghurt, concentrates, juices, syrups and wines. Strawberries are processed into jams and jellies by heating the fruit and then adding pectin and sugar. Fruit yoghurt is made by adding jam to natural yogurt. Hot syrup is poured over the fruit in canning.

Most processed strawberries are hand-picked in the field without their caps (receptacles) and put in large containers. Machines are also available to mechanically harvest fruit (see Chapter 6), but these have not become popular. Premium price is paid for strawberries that are of uniform size, firm, red throughout, and have high soluble solid content and flavour. Varieties vary greatly in their processing quality, with 'Hood' in the Pacific Northwest of the US and 'Senga Sengana' in Europe being among the most favoured varieties in the world.

Most commonly, fruit are individually quick frozen (IQF). This product is then distributed as it is, or used for other products. The advantage of this procedure is that such fruit can be removed individually from the container or package. For IQF, fruit are usually frozen in a blast of liquid CO_2 or are immersed in liquid nitrogen. Liquid CO_2 freezes the fruit in 11 min whereas liquid nitrogen results in frozen fruit in 8 s. Strawberries are also frozen in bed freezers with blasts of cold air that lift the fruit and cause them to flow towards their final containers (Gruda and Kurzeba, 1981). The liquid nitrogen and CO_2 freezing methods are superior to air blast freezers as they freeze the fruit much faster, resulting in far less water loss compared with air blast freezers, and soft or wet fruit are less likely to be frozen together in chunks (Smith *et al.*, 1974; Lucas, 1979). Immersion in liquid nitrogen freezes fruit the fastest, but can result in the fruit being shattered if they are immersed too long.

Strawberries are also frozen in bulk with sugar (one part sugar to four to five parts fruit). The addition of sucrose often enhances the flavour of a frozen pack, helps to minimize ice crystal formation and helps to stabilize colour (Hudson *et al.*, 1977). Sucrose from sugar beet or cane sugar is the sweetener used most commonly commercially. High fructose corn syrup, can also be used, but its overpowering sweetness is less preferable (Perkins-Veazie and Collins, 1995).

Freezing and thawing greatly alters strawberry fruit quality. In general, frozen fruit have reduced colour and odour, are less sweet and more sour than fresh fruit (Fig. 7.5; Douillard and Guichard, 1990; Schreier, 1980).

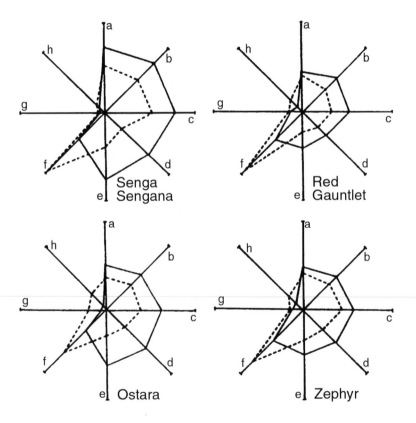

Fig. 7.5. Differences in quality between frozen (dashed lines) and fresh fruit (solid lines). The axes are: (a) intensity of odour; (b) character of odour; (c) overall impression of odour; (d) sweetness; (e) taste; (f) sourness; (g) off-odours; (h) off-tastes (redrawn from Hirvi, 1983).

Strawberry storage at −40 or −80°C is superior to −20°C as flavour esters are lost more slowly, whereas soluble solids content (SSC) and total acids (TA) do not change (Deng and Ueda, 1993). Frozen fruit flavour depends largely on the ratio of SSC to TA, with bland fruit having high SSC and low TA, and highly flavoured fruit having high SSC and high TA (Kader, 1991).

Fruit firmness is also reduced during freezing and thawing. Fruit must be firm before they are frozen to be firm afterwards, and precooling to −1°C greatly improves the percentage of usable frozen fruit (Conroy and Ellis, 1981). Preprocessing dips of calcium, pectin and sucrose have been found to have significant effects on the firmness, drained weight loss and colour of frozen-then-thawed fruit and preserves (Morris *et al.*, 1985b, 1991). Sliced fruit generally are firmed more than whole fruit.

Loss of red pigmentation and subsequent browning is one of the primary concerns during all types of processing including frozen and thawed berries (Sistrunk and Cash, 1968; Wrolstad *et al.*, 1970), preserves (Sistrunk and Morris, 1978; Abers and Wrolstad, 1979), low-sugar spreads and fillings (Sistrunk *et al.*, 1982; Pratt *et al.*, 1986), and concentrates and juices (Wrolstadt *et al.*, 1990; Rwabahizi and Wrolstad, 1988). Anthocyanin in processed fruit and juices degrades rapidly due to phenolic polymerization and oxidative degradation of pigments (Wrolstad *et al.*, 1990; Skrede *et al.*, 1992; Deng and Ueda, 1993). Polyphenol oxidase indirectly degrades anthocyanins by forming quinones and other intermediary compounds through the oxidation of catechin (Wesche-Ebeling and Montgomery, 1990). Likewise, peroxidase isozymes have been implicated in the indirect degradation of anthocyanins (López-Serrano and Barceló, 1996). Colour stability in processed fruit is greatly enhanced by high initial anthocyanin content, low pH, high sugar content, cold temperatures and high levels of antioxidants (Sistrunk and Cash, 1968; Wrolstad *et al.*, 1970, 1990). Limiting mould contamination is also critical, particularly with juices and concentrates (Rwabahizi and Wrolstad, 1988; Lundahl *et al.*, 1989).

CONCLUSIONS

Strawberry flavour is a complex combination of sweetness, acidity and aroma. The main soluble sugars are glucose and fructose, and the primary organic acid is citric acid. Hundreds of volatile esters are produced during strawberry ripening, with methyl and ethyl esters of butanoic and hexanoic acids being the most prevalent. Our knowledge about the specific components of flavour are only beginning to emerge, but furaneol and furanone are thought to be particularly important components of strawberry aroma.

Strawberries are not climacteric, but strawberry ripening is associated with numerous biochemical changes including increases in pectins, hemi-celluloses and several enzymes associated with anthocyanin and fatty acid biosynthesis. Red colour develops primarily through the production of the anthocyanin pelargonidin-3-glucosidase, although cyanidin 3-glucoside increases in importance during storage. Soluble solids content rises continuously during fruit development from 5% in green fruit to as much as 9% in ripe ones. The pH of strawberry remains relatively stable during fruit development at about 3.5, but titratable acidity gradually drops.

Strawberry appearance and fruit quality is dependent on a number of pre- and postharvest factors. Fruit shape is regulated to a large extent by the number of fertile achenes per fruit and the resulting patterns of auxin production. For the fresh market, fruit that are three-quarters ripe develop adequate colour and store longer than those picked fully ripe, but the best fruit for the processed market are those fully coloured. Elevated temperatures can have a negative effect on fruit size and firmness during fruit development, and the rapid removal of field heat after harvest is critical to preserving fruit quality. Postharvest quality of fresh fruit can also be extended by controlled atmosphere storage at elevated CO_2 (10–20%) and decreased O_2 (5–10%). Other methods of extending storage life, such as modified atmosphere packages or irradiation, have not been widely implemented, but a number of natural volatilization compounds show high promise as a means to control rots.

8

PEST AND DISEASE MANAGEMENT

INTRODUCTION

Numerous diseases and pests cause serious economic damage to strawberries across the world (Schaefers, 1990; Wilhelm and Nelson, 1990; Maas, 1998). All parts of the plant are attacked by a broad array of insect, fungal, virus and nematode pests (Tables 8.1 and 8.2). This chapter describes the biology of some of the most important diseases and insects, and outlines management strategies. Emphasis is placed on reviewing integrated crop management strategies that incorporate cultural, genetic and biological control methods (Kovac et al., 1993; Bostanian et al., 1994; Cooley and Schloemann, 1994; Strand, 1994).

Table 8.1. Common insect pests of strawberries.

Plant part affected	Symptoms	Pest	Appearance
Flowers	Dangling, damaged buds	Bud weevil	Dark, reddish-brown weevil about 2.5 mm long, with a slender, curved snout about half as long as the body; two large black spots on the body
Flowers and fruits	Small seedy fruit with dull or bronzed colour	Flower thrips	Slender, winged adults that are orange or yellow; about 1 mm in length
	Small berries with concentration of seeds in centre	Tarnished plant bugs	Adult tarnished plant bugs are about 6 mm long, brownish in colour with yellowish and black dashes; overall brassy appearance

Table 8.1. *Continued*

Plant part affected	Symptoms	Pest	Appearance
Fruits	Deep holes in berries	Sapbeetles	Small, oval beetle less than 3 mm long; brown with a slightly mottled appearance
Leaves and fruits	Folded leaves with silken threads; tunnels in berries	Omnivorous leaf tier	Moths (10–12 mm long) have buff-brown or grey wings; young larvae have white bodies and shiny black heads; nearly mature larvae have three distinct light grey stripes and their heads are light brown with dark spots on either side
	Crinkled leaves with short leaf stems; retarded plant growth and flat appearance; small, dry and distorted fruit	Cyclamen mites	Barely visible with the naked eye; white to caramel in colour as an adult; milky white in immature stages
Leaves	Leaves covered with honeydew; stunted plants	Aphids	Small, soft-bodied translucent insects; less than 1.6 mm long; winged and wingless forms
	Folded leaves with silken threads	Leafrollers	Adults are reddish-brown moths with a wing span of over 10 mm; young larvae are pale green changing to grey–brown as they mature; pupae are yellowish brown
	Small chlorotic spots on expanding leaves	Mealybugs	Soft-bodied insects (2 mm long) that are covered with a loose deposit of wax; have a cottony white appearance
	Curled distorted leaves with interveinal yellowing	Potato leafhopper	Adults are brownish to green and about 3 mm long; nymphs are light green and move sideways when disturbed
	Mottling, speckling and bronzing of foliage; tangle of fine webbing on undersides of leaves	Spider mites	The adult has eight legs and is tiny (0.5 mm). Pale greenish to yellow, with two dark spots

Table 8.1. *Continued*

Plant part affected	Symptoms	Pest	Appearance
Leaves	Frothy, irregular masses of foam	Spittlebugs or froghoppers	Nymphs are small, orange to green insects
	Sticky undersides of leaves	White flies	Adult is a small (1.5 mm), white four-winged insect
Leaves and roots	Larvae feeding on fine roots	Grape colaspis	The adult is about 5 mm long; it is yellowish brown with brown legs and yellowish brown antennae
	Leaves riddled with small holes; larvae feeding on roots	Rootworms	Adult is a shiny, oval beetle, brown to black with four dark blotches on the wing covers; about 3 mm long
Roots and leaves	Low vigour; pale foliage and desiccated fruit	Root aphid	Bluish green nymphs
Roots	Stunted plants; root feeding	Garden symphilid or centipede	Adults are white, about 7 mm long, with a pair of long-beaded antennae; nymphs start with six legs and add one pair each moult until there are 12
	Red leaves and undersized berries, particularly during drought	Root weevils	Black flightless adults, range from 5 to 10 mm long with pronounced nose or snout; legless C-shaped grubs
	Low vigour plants; root feeding	White grubs	Grubs are large, thick bodied and dirty white; up to 25 mm long
Crowns	Tunnelling in crowns; weak, red plants	Crown borer	Larvae are white, legless grubs about 5 mm long; adult is a chestnut brown, short snouted beetle with three irregular dark spots on each wing cover
	Tunnelling in crowns; plants with yellow centres	Crown miner	Reddish pink larvae that are 12–12.5 cm long
	Hollowed out crowns, dead crowns or plants	Crown moth	Adults are predominantly black (12–13 mm long), with distinct yellow stripes on their bodies; larvae have brown heads and whitish or pinkish, distinctly segmented bodies

Table 8.2. Common bacterial and fungal diseases of strawberry.

Plant part affected	Symptoms	Disease
Leaves	Water soaked lesions on undersides of leaves that develop into large angular spots; translucent in transmitted light	Angular leaf spot
	Dark purple to reddish purple, round spots on uppersides of leaves; centres of spots turn tan and then white	Leaf spot
	Purple to reddish purple, angular spots on uppersides of leaves; centres of spots remain purple	Leaf scorch (red spot)
	Reddish purple to brown spots that spread into V shapes encompassing leaf margins	Leaf blight
	Deep purple, round to irregularly shaped spots on upper leaves; brown centres that turn white or rusty brown; spots often coalesce to form 'scorched' areas	Ramularia leaf spot
	Curling of leaves: dry purplish brown patches on undersides of leaves; reddish discoloration on upper surfaces	Powdery mildew
Fruit	Begins as soft light brown areas; fruit become mummified with a grey, dusty powder covering them	Grey mould (ash mould)
	Begins as dark brown or natural green areas with brown margins in green fruit; the whole berry becomes brown and leathery; fruit taste bitter	Leather rot
	Begins as white water soaked lesions that become brown and eventually encompass the whole fruit; lesions contain salmon-coloured spore masses	Anthracnose
Roots and crowns	Young leaves suddenly wilt, become necrotic brown and the whole plant rapidly dies; crowns are reddish brown and break easily at their top	Cactorum crown rot or vascular collapse
	Leaves rapidly wilt and the whole plant dies; crowns turn brown	Anthracnose
	Leaves rapidly wilt, become reddish-yellow or dark brown, and dry; new roots dwarfed with blackened tips	Verticillium wilt
	Plants are stunted ; young leaves are metallic bluish-green, old leaves turn yellow or red; the stele of roots is pink to brick red	Red stele
	Plants are stunted with small root systems; main roots spotted with dark patches or black; few feeder roots	Black root rot

ARTHROPOD AND MOLLUSC PESTS

Piercing and sucking insects

Aphids

Aphids are found in every strawberry growing region (Williams and Rings, 1980; Antonelli *et al.*, 1991; Maas, 1998). There are several strawberry aphids of widespread importance that attack primarily leaves: *Chaetosiphon fragaefolii* (Cockerell), *C. thomasi* (H.R.L.), *C. minor* Forbes and *C. jacobi* H.R.L. The strawberry root aphid, *Aphis forbesi* Weed, feeds on leaf buds, leaves and roots. Aphids are of primary concern as vectors of virus diseases, but heavy feeding can result in stunted plants. Winged and wingless aphids are most numerous in the spring and autumn, but several generations are produced each year.

Spider mites

Spider mites are a worldwide problem (Antonelli *et al.*, 1991; Strand, 1994; Maas, 1998). The most troublesome species are *Tetranychus urticae* Koch. (two-spotted spider mite), *T. telarius* L. (northern or green two-spotted spider mite or red spider mite) (Fig. 8.1), *T. lobustus* Boudreau (southern lobed mite or red spider mite), *T. turkestani* Ugarov and Nikolski *(T. atlanticus)* (strawberry spider mite) and *T. cinnabarinus* (Boisduval).

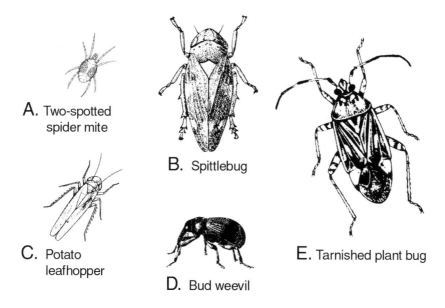

A. Two-spotted spider mite

B. Spittlebug

C. Potato leafhopper

D. Bud weevil

E. Tarnished plant bug

Fig. 8.1. Common piercing and sucking insects found on strawberries (drawings from Funt *et al.*, 1997, with permission).

Both adults and immatures feed by rasping and sucking leaf surfaces, which causes mottling, speckling and bronzing of foliage. They commonly cover the undersides of leaves with a fine webbing. Spider mites overwinter in all kinds of vegetation around strawberry fields and become active in the spring. Many generations are produced each year. The mites are wingless, but can float long distances on the silken threads they produce. Drought stressed and unthrifty plants are most subject to attack.

Lygus or tarnished plant bugs

These are widespread throughout North America, Europe and Asia (Bostanian, 1994; Strand, 1994; Pritts and Handley, 1998). Three species cause extensive damage to strawberry fruit including *Lygus lineolaris* (Palisot de Beauvois), *L. hesperus* Knight and *L. elisus* Van Duzee (Fig. 8.1). Deformed berries are caused by the feeding activity of both adults and nymphs on flower buds and developing fruit. Tarnished plant bugs suck sap from developing seeds and receptacles, killing surrounding cells (Handley and Pollard, 1993). Ripening berries often remain small with a concentration of seeds in their centre.

Adults overwinter in vegetation or debris, and in the spring, they are attracted to the flower buds and shoot tips of a wide array of plants, including many tree and small fruit crops. Strawberries are the preferred host in the eastern US and Canada, but in California, they prefer local weeds until these weeds begin to dry up in the spring.

Potato leafhopper

The potato leafhopper, *Empoasca fabae* (Harris), reduces plant growth and runner production throughout eastern North America (Fig. 8.1). They feed primarily on the undersides of leaves, leaving them curled, distorted with a yellow colouring between veins (Williams and Rings, 1980; Maas, 1998). They are most active in late summer and early spring.

Spittlebugs or froghoppers

Spittlebugs or froghoppers, *Philaenus spumarius* (Linnaeus) (Fig. 8.1) are common across most of the eastern United States and along the Pacific coast (Williams and Rings, 1980; Antonelli *et al.*, 1991; Maas, 1998). Probably best known for their unsightly foam, they can reduce plant vigour and yield by sucking plant sap. Spittlebugs overwinter as egg masses in strawberry fields, small grains or lucerne. They emerge about the same time as strawberry flowers begin to appear. The nymph stage lasts for 5–8 weeks, with the spittle being produced as soon as the nymphs begin eating. They start feeding at the base of plants and move up as the season progresses. There is only one generation per year.

Strawberry bud weevil

The strawberry bud weevil (clipper), *Anthonomus signatus* Say (Fig. 8.1), is an important pest in the eastern United States and Canada (Williams and Rings, 1980; Mailloux and Bostanian, 1993; Maas, 1998). The adults puncture buds with their snouts to feed on immature pollen. Females deposit a single egg within buds, girdle them and then clip the stems. The damaged buds dangle loosely or fall to the ground. Some adult clippers overwinter in strawberry fields, but most are found in nearby fences and woodlands. They emerge about the same time that strawberries begin flowering.

Other insect pests

Two other common greenhouse pests occasionally cause damage to strawberry plantings (Maas, 1998). The cyclamen mite, *Phytonemus pallidus* (Banks), feeds on young unfolding leaves and on blossoms, resulting in distorted, purplish leaves and malformed blossoms. White flies, *Traialeurodes packardi*, can be found on the undersides of leaves sucking plant sap, particularly in greenhouses. They produce a sticky honeydew.

Chewing insects

Flower thrips

Flower thrips, *Frankliniella* spp. (Fig. 8.2), are a worldwide problem in both open and protected culture (Maas, 1998; Pritts and Handley, 1998). Thrips feed on developing seeds and the tissue between seeds. The resulting damaged fruit are small, seedy and have a dull or bronzed colour. The fruit may also develop unevenly. In North America, flower thrips overwinter in the south and migrate north on high-level winds associated with weather fronts (Funt *et al.*, 1997). They deposit eggs in plant tissue and have two active nymph stages and two pupal-like nymph stages. Their life cycle is completed in several weeks, with several generations per year.

Root weevils

Over 20 species of root weevil (Fig. 8.2) attack strawberries (Strand, 1994; Maas, 1998). Probably the most important are *Otiorhynchus sulcatus* (Fabricius) and *O. ovatus* (Linnaeus) with worldwide distributions, and *O. meridionalis* Gyllenhal, an occasional field pest in California that has caused severe damage in France under polytunnels.

The larvae of root weevils feed on strawberry crowns and roots (Cram and Neilson, 1978; Strand, 1994; Pritts and Handley, 1998). The worst damage occurs when plants are under periods of stress. Infestations generally occur in patches in the field, and affected plants have red leaves and undersized berries. Adult weevils also chew notches on the edges of leaves at night, but this damage is rarely significant. Adults cannot fly.

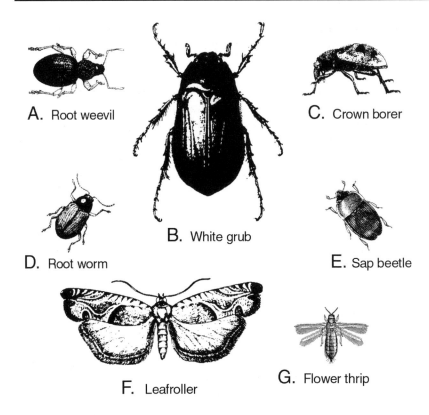

A. Root weevil

C. Crown borer

B. White grub

D. Root worm

E. Sap beetle

G. Flower thrip

F. Leafroller

Fig. 8.2. Common chewing and boring insects found on strawberries (drawings from Funt *et al.*, 1997, with permission).

Root weevils overwinter as full grown larvae, pupae or adults in the soil, or as adults in plant debris. Larvae and pupae complete development in late spring and together with the adults become active in May. Root weevil adults lay eggs in the soil of strawberry fields throughout the summer, with single females laying 150–200 eggs. The eggs hatch in about 10 days and the resulting larvae burrow into the soil to feed on roots until they mature.

White grubs

White grubs (Fig. 8.2) are commonly the larvae of species of *Phyllophaga* spp. (June bugs or May beetles), *Cyclocephala borealis* Arrow (northern mask chafer) or *Popillia japonica* Newman (Japanese beetle) (Maas, 1998; Pritts and Handley, 1998). The risk of infestation is highest in new plantings established on former sod or grass fields (Funt *et al.*, 1997). Females deposit eggs in soil during late spring or early summer, eggs hatch in 2 or 3 weeks, feed on roots and then burrow deeply into the soil to overwinter. After 3 years, the adults emerge in the spring and feed on leaves leaving them skeletonized. Adults hide in the soil during the day and feed on leaves at night.

Leafrollers

There are dozens of leafrollers species that damage strawberries (Cram and Neilson, 1978; Antonelli *et al.*, 1991; Strand, 1994). Among the most important are: *Ancylis comptana fragariae* (Walsh & Riley) (strawberry leafroller), *Choristoneura rosaceana* Harris (oblique-banded leafroller), *Compsolechia fragariella* (western strawberry leafroller), *Ptycholoma peritana* (garden tortix) and *Cnephasia longana* (omnivorous leaf tier) (Maas, 1998). The strawberry leafroller (Fig. 8.2) has the widest distribution, and was introduced to the USA from Europe in the last century. The garden tortix is the most common leafroller pest in California (Strand, 1994).

Leafrollers fold and tie together leaves with silken threads. They feed solely on the epidermis layers, but the whole leaf eventually turns brown. Leaf-rollers overwinter as larvae or pupae in folded leaves or litter. They emerge in the early spring and deposit translucent eggs on the lower surface of leaves. The eggs hatch in 1 or 2 weeks and pupate for about a week in folded leaves. Several generations can be produced per season.

Strawberry rootworm

Strawberry rootworm, *Paria fragariae* Wilcox (Fig. 8.2) is a widespread pest (Strand, 1994; Maas, 1998; Pritts and Handley, 1998). In some locations, it is most damaging as an adult leaf feeder, whereas in other areas it most severely damages roots as larvae. They overwinter in mulch and soil crevices, and become active in May and June. Adults chew small holes in leaves at night, and females lay their eggs in leaves near the soil surface. The larvae burrow through the soil to roots, where they feed until mid-summer when they begin to emerge and feed on leaves. Heavy infestations result in plants having leaves riddled with small holes.

Sap beetles

Strawberry sap beetles, *Stelidota geminata* (Say) (Fig. 8.2) chew deep cavities in ripe berries that are unsightly and can lead to infection by rot organisms (Williams and Rings, 1980; Pritts and Handley, 1998). It is probably the most damaging pest attacking ripe fruit in North America (Maas, 1998). Adults fly into fields from wooded areas as the fruit begin to ripen. They deposit eggs in ripe berries that hatch in 2 to 3 days.

Other chewing insects

Other important chewing insects include the following (Maas, 1998): (i) the garden symphilid, *Scutigerella immaculata* (Newport), becomes important in the Pacific Coast of North America only when large populations have built up on other crops such as asparagus. They feed on roots. (ii) The omnivorous leaf tier, *Cnephasia longana* (Haworth) is a limited pest in California and the Pacific Northwest, but can be extremely troublesome, because its larvae bore into fruit and remain inside. (iii) Grape colaspis, *Colaspis brunnea* (Fabricius), is primarily a pest of grapes, but can cause serious damage in the eastern USA. Its larvae feed on small roots and the adults feed on leaves, leaving many holes. Unlike the strawberry rootworm, grape colaspis feed during the day. (iv) Rose chafers, *Macrodactylus subspinosus*, skeletonize leaves as adults.

Boring insects

Strawberry crown borer

The strawberry crown borer, *Tyloderma fragariae* (Riley) is widespread in North America except at high elevations (Maas, 1998). Strawberry plants are weakened and turn red or are killed as the larvae bore downward into the crowns. Infestations usually begin at field borders or near older portions of fields that are heavily infested. Spread is gradual. The adults, which cannot fly, chew many small, round holes in leaves in the autumn, but this damage is rarely critical. Adults overwinter in plant debris in strawberry fields or in protected areas nearby. They become active at about the same time as bloom, by opening holes in crowns where they lay eggs that develop in about a week. Eggs hatch into pupae that feed for several weeks before emerging as adults in the autumn. Egg laying continues until early summer.

Strawberry crown miner

Strawberry crown miner, *Monochroa fragariae* (Busck), are serious pests in the Pacific Northwest and in parts of the midwestern US (Antonelli *et al.*, 1991; Maas, 1998). Infested plants develop yellow centres in the autumn and often die. The crown miner overwinters in crowns as fully grown larvae. They pupate in the early spring and begin laying eggs in the latter part of June for about 2 weeks. Crown miners lay their whitish eggs on the sheaths surrounding the crown, petioles and the undersides of leaves. When the eggs hatch, the larvae drop to the ground and bore into the sides of crowns, slightly below the leaves. Larvae survive the winter in silken cases at the base of the crowns.

Strawberry crown moth

Strawberry crown moth, *Synanthedon bibionipennis* (Boisduval), is found along the Pacific coast of North America. Although damage is generally restricted to a few individuals in a field, some major infestations do occur (Maas, 1998). The larvae hollow out crowns, killing the plants. Moths emerge in early summer and lay flat, brown eggs singly near plant crowns. Eggs hatch in about 10 days, the emerging larvae bore into crowns, tunnel extensively and then generate silken cocoons for overwintering. They feed on crowns for another season, spin a second cocoon and then pupate the following winter.

Slugs

There are a number of slug pests on strawberries in *Deroceras* and *Arion*, but the Arion slug, *Arion subfuscus*, is probably the most vigorous, being active even in bright sunlight (Maas, 1984; Pritts and Handley, 1998). Slugs damage fruit by eating deep ragged holes on berries, primarily under the receptacle. They also leave slime trails as they move around. Slugs hatch from eggs deposited the previous autumn in the soil of strawberry plantings. Moist, thick straw mulches favour their development.

Control of insect pests

The key to control of most of the insect pests is good sanitation and isolation (Table 8.3). The fruit sucking and boring insects are controlled to a large extent by removing damaged and overripe fruit from the field. The perennial damage of leaf feeders is reduced by renovating fields each year. Damage from slugs is minimized by proper plant spacing leading to good air circulation, and removing mulch after harvest. Adults of crown borers and root weevils can only fly short distances and as a result damage to new plantings can be minimized by isolation from other infested strawberry fields and woods. Lygus bugs invade early flowering weeds so weed control both within and between fields helps to diminish infestations. Many of the root and crown feeders have adults stages that are difficult to control, but future damage can be greatly diminished by destroying infested areas by tillage and avoiding diseased areas when transplanting.

Chemical controls have been developed for most of the common pests (Schaefers, 1990; Strand, 1994; Funt *et al.*, 1997), but management strategies that rely solely on regular applications of broad-spectrum insecticides often waste chemicals, encourage the evolution of resistance in the pests and reduce populations of natural predators (Strand, 1994; Easterbrook *et al.*, 1997). Some of the more important beneficial insects that are injured by insecticides include: (i) bigeyed bugs (*Geocoris* spp.), which feed on mites, lygus bugs, aphids and caterpillar pests; (ii) lacewings (*Chrysoperla* and *Chrysopa* species) which feed on anything they can capture, as well as insect eggs and mites (Fig. 8.3); (iii) minute pirate bugs (*Orius* spp.) which prefer thrips; (iv) lady beetles (*Hippodamia* spp.) which eat aphids (Fig. 8.3); (v) predatory mites (numerous species) which feed on two-spotted spider mites; and (iv) parasitic wasps (*Anaphes iole*) which lay their eggs in lygus bug eggs (Fig. 8.3).

To minimize chemical use, integrated crop management (ICM) strategies have been developed for many strawberry pests. Regular monitoring is done of pest numbers and pest damage, weather data and crop development (Cooley and Schloemann, 1994; Strand, 1994; Maas, 1998). Pesticide applications are delayed until pest populations are approaching a threshold size where real

Table 8.3. Summary of integrated crop management procedures currently utilized in strawberries.

I. Cultural controls when establishing a new planting
    A. Site selection:
        1. Site new plantings on land with well-drained soil.
        2. If planting on a site previously in sod or grass, evaluate presence and density of white grubs.
        3. Avoid siting a planting near woods or fencerows to avoid migration of clipper and root weevils.
        4. Avoid siting a new planting near old plantings to avoid infestation of aphids, clipper, crown borers, fungal diseases, miners, moths and root weevils.
    B. Site preparation
        1. Plant cover crop(s) during the year prior to strawberry planting.
        2. Adjust soil pH to 5.5–6.5.
        3. Cultivate deeply to break up any hard pan or other subsoil problems.
    C. Culitivar selection:
        1. Select clean plant material from a reputable nursery to avoid problems with angular leaf spot, crown rot, cyclamen mite, leaf fungal diseases and viruses.
        2. Plant varieties resistant to insect pests, leaf diseases, root rots and crown rot.
    D. Plant establishment
        1. Use a raised bed planting system for sites with excessive soil water or poor drainage to minimize fungal diseases of roots and fruit.
        2. Establish plant rows with adequate spacing to minimize fungal diseases of leaves and fruit.

II. Cultural controls in established plantings:
    A. Keep weeds under control both in and between fields to avoid problems with spittle bugs and tarnished plant bugs.
    B. Promptly remove all ripe fruit and cull berries to minimize sap beetle and fruit rot damage.
    C. Maintain mulch between rows to suppress weeds and control fruit rots.
    D. Subsoil between rows to control root aphids.
    E. Renovate as soon after harvest as possible
        1. Map existing weed problems for spot control.
        2. Mow foliage to help manage clipper, slugs, mites and fungal leaf diseases.
        3. Till in straw and leaf litter between rows to minimize damage from fungal diseases of fruit and leaves.
        4. Subsoil between rows if soil has become compacted to help control black root rot.
        5. Control weeds as needed with pre-emergence herbicides
    F. Rotate every 2–4 years for 2–3 years with non-host crops to minimize problems with various pests especially root pathogens and weeds.
    G. Avoid standing water and saturated soils to minimize spread of fungal and bacterial diseases.

Table 8.3. *Continued*

    H. Remove litter in and around fields to minimize damage from leafrollers, root worms and fungal diseases of fruit and leaves.

    I. Destroy areas infested with root and crown feeders to diminish spread of these pests.

III. Biological and mechanical control options:
    A. Release predator mites to control two-spotted spider mite.
    B. Vacuum two-spotted spider mites.
    C. Use milky spore to control white grubs.

IV. Scouting for pests (known thresholds are in parentheses):
    A. Pre-bloom (once a week).
        1. Look for dangling buds damaged by strawberry clipper (five clipped buds per 0.3 m linear row).
        2. Examine leaflets for two-spotted spider mite (5–20 mites per mid-tier leaflet in susceptible cultivars).
        3. Sweep net sample for tarnished plant bug adults (two adults per 10 sweeps).
        4. Check aphid populations (about 30 aphids per plant).
        5. Look for strawberry rootworm beetle feeding injury (shot-holes) on leaves (no known threshold).
    B. During bloom and into harvest (once per week).
        1. Examine flowers for flower thrips (0.5 thrips per fruit).
        2. Examine flowers for tarnished plant bug nymphs (0.25 nymphs per flower or 10% infested).
        3. Examine plant stems for spittlebug (more than one spittle mass per metre row).
        4. Examine leaflets for two-spotted spider mite (5–20 mites per mid-tier leaflet in susceptible cultivars).
        5. Check aphid populations (about 30 aphids per plant).
        6. Look for rootweevil feeding (notching of leaf margins) on foliage (no known thresholds).
    C. Post-harvest (once every 2 weeks).
        1. Examine leaflets for spider mites (5–20 mites per mid-tier leaflet in susceptible cultivars).
        2. Look for strawberry rootworm beetle feeding damage (shot holes) on leaves (no known thresholds).
        3. Look for leafrollers, white flies and aphids.
        4. Map weed infestations.

economic losses occur (Table 8.3). The most extensive ICM programmes have been developed for two-spotted spider mites and lygus bugs (Bostanian *et al.*, 1994; Cooley and Schloemann, 1994; Strand, 1994).

Numerous promising biological controls have been identified that can be incorporated into ICM programmes. The green lacewing *Mallada basalis* has been shown to effectively control two-spotted spider mite (Chang and Huang,

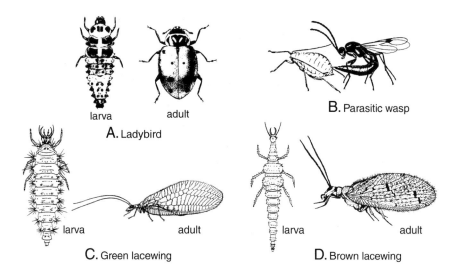

Fig. 8.3. Common beneficial insects found on strawberries (drawings from Funt *et al.*, 1997, with permission).

1995), together with a number of predatory mites including *Amblyseius fallacis, Metaseiulus occidentalis* and *Phytoseiulus macropolis* (Maas, 1998), *Phytoseiulus persimilis* (Trumble and Morse, 1993; van de Vrie and Price, 1994 ; Easterbrook *et al.*, 1997; Price and Nguyen, 1997), *Neoseuilus fallacis* (Coop and Croft, 1995) and *Typhlodromus pyri* (Zacharda and Hluchy, 1996). To date, the predatory mite, *Phytoseiulus persimilis*, has had the most wide-spread commercial use (Strand, 1994).

A number of mycopesticides and nymphal parasitoids have been identi-fied as controlling *Lygus* bugs (Hedlund and Graham, 1987; Sohati *et al.*, 1992), such as *Beauveria bassiana* (Maas, 1998), *Peristenus digoneutis* (Day *et al.*, 1990), *Leiophron uniformis* (Norton *et al.*, 1992) and *Anaphes iole* (Jones and Jackson, 1990; Norton and Welter, 1995, 1996). Vacuums have also been developed to remove lygus bugs from fields (Zalom *et al.*, 1993; Pickel *et al.*, 1995).

Among the chewing and boring insects, black vine beetle larvae have been sucessfully controlled by soil treatment with the nematodes *Heteror-habditis* sp. (Backhaus, 1994) and *Steinernema carpocapsae* (Sampson, 1994; Kakouli Duarte *et al.*, 1997). Numerous parasitoid species have been re-covered from strawberry leafroller larvae and pupae that could be used in control strategies (Obrycki *et al.*, 1993). *Amblyseius* spp. have shown promise in the field control of thrips (Wardlow, 1994) and it has been shown that the principal predators of thrips can be maintained at high levels by planting broad beans adjacent to strawberries (González-Zamora *et al.*,1994). Several entopathogenic fungi and nematodes have shown potential in controlling the root weevils, *Otiorhynchus ovatus* and *O. dubius* (Vainio and Hokkanen, 1993), as well as carabid beetles (Maas, 1998). The parasitic wasps, *Microctonus nitidulidus* and *Brachyserphus abruptus* are known to attack strawberry sap beetles (Maas, 1998). In glasshouses and plastic tunnels in Belgium, thrips have been controlled by *Amblyseius cucumeris* and *Orius insidiosus*, and *Amblyseius californicus* and *Therodiplosis persicae* have worked well against spider mites (Sterk and Meesters, 1997). In addition, aphids have been successfully controlled by *Aphelinus abdominalis*, *Aphidius matricariae* and *Aphidius ervi*, *Hippodamia convergens* and three species of *Hymenoptera*.

COMMON STRAWBERRY DISEASES

Bacterial diseases

Angular leaf spot, *Xanthomonas fragariae* Kennedy & King, is a rapidly growing worldwide problem in strawberries (Maas *et al.*, 1995). It typically starts as minute, water-soaked lesions on the undersides of leaves and then develops into large angular spots that are delineated by small veins. They appear dark green in reflected light, but are translucent in transmitted light. Lesions grow in size and coalesce until they are difficult to distinguish from leaf spot and scorch, except that under moist conditions, the lesions produce a viscous exudate that dries into a whitish, scaly covering. The pathogen overwinters in dead leaves that are dry or buried in the soil. It enters hosts' tissue passively through wounds or actively as mobile cells that swim into natural plant openings. Disease development is favoured by moderate to cool day temperatures (20°C), night temperatures near zero and high relative humidity and precipitation (including frost control). Healthy plants are more prone to attack than sick and diseased ones.

Control of bacterial diseases

Antibiotics and copper-containing compounds provide some protection against *Xanthomonas fragariae*, but the number of copper applications necessary for complete control may be phytotoxic (Maas *et al.*, 1995). Disease avoidance may be the key to control through planting of healthy, resistant planting stock and eradication of affected plantings (Milholland, 1996). Genetic analyses of isolates from diverse locations showed little differences among populations, suggesting that this disease is spread primarily through the exchange of infected plant material (Pooler *et al.*, 1996).

Fungal diseases of the foliage

Leaf spot

The leaf spot (ramularia leaf spot) organism, *Mycosphaerella fragariae* (Tul.) Lindau, infects leaves, fruits, petioles, runners, fruit stalks and berry calyxes. It is probably the most widespread leaf disease of strawberries, with several physiological races being identified (Maas, 1998). The most common symptom is the appearance of dark purple to reddish purple spots (5–10 mm wide) which develop on the upper surfaces of leaves (Fig. 8.4), although symptom expression is dependent on variety, temperature regime and strain (Nemec, 1971; Maas, 1998). The centres of these spots turn tan or grey and eventually white. Late in the season, tan or bluish areas form on leaf undersurfaces. One or two black spots (black seed disease) may form on fruit in particularly warm weather. Older infected leaves that survive the winter give rise to conidia that are spread to new foliage by splashing water or handling infected plants. Perithecia form at the edges of leaf spots in the autumn and ascospores are ejected in the spring. These are carried to new leaf tissue by wind and water.

Leaf scorch

Early symptoms of leaf scorch (red spot), *Diplocarpon earliana* (Ellis & Everh.) Wolf , are very similar to the early stages of leaf spot, but the spots become angular or irregular and the centres of the spots remain dark purple (Converse *et al.*, 1981; Nemec and Blake, 1971) (Fig. 8.4). The entire leaf may become reddish or light purple. The leaf scorch fungus infects leaves, fruits, petioles, runners, fruit stalks and berry calyxes. The fungus overwinters on infected leaves and conidia are produced in black acervuli. Ascospores are also produced in the early spring within disc-shaped, black apothecia. Conidia are produced throughout the season when moisture levels are adequate; these conidia are spread mostly by splashing water (Zheng and Sutton, 1994). Leaf scorch occurs throughout the range of strawberry cultivation.

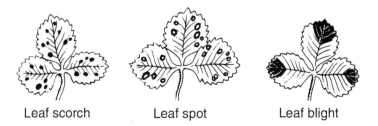

Leaf scorch   Leaf spot   Leaf blight

Fig. 8.4. Common strawberry leaf diseases (drawings from Funt *et al.*, 1997, with permission).

Leaf blight

Leaf blight, *Phomopsis obscurans* (Ellis & Everh.) Sutton, is most common on plants after harvest (Eshenaur and Milholland, 1989). It is distinct from leaf spot and leaf scorch, in that, the infected area becomes V-shaped with the widest part of the V at the leaf margin (Fig. 8.4). Spots are initially reddish purple and they develop a darker brown or reddish-brown centre surrounded by a light brown area with a purple border. Leaves can eventually become blighted and die. This fungus also infects fruit caps with similar symptoms, and it can cause a soft, pale pink rot at the end of fruits (Maas, 1985). The fungus overwinters on both dead and living tissue. Conidia are produced in black pycnidia embeded in the centres of old lesions. Conidia are spread primarily by water.

Powdery mildew

Powdery mildew is caused by the fungus *Sphaerotheca macularis* (Wallr.: Fr.) Jacz. f. sp. *fragariae* Peries in all parts of the world (Peries, 1962; Spencer, 1978). The first symptoms are generally a curling of leaf edges, followed by dry purplish to brownish patches that develop on the lower surface of leaves. As the disease progresses, patches of white, powdery fungus mycelium develop on the undersides of leaves and a reddish discoloration appears on the upper surface. Infected fruit become hard and fail to ripen normally. The pathogen overwinters on infected leaves and the spores are carried by wind to new growth in the spring. Spread is facilitated by temperatures of 15–27°C and moderate to high humidity.

Other fungal diseases

Numerous other pathogens cause localized, economic problems across the world (Maas, 1984). Alternaria black leaf spot, caused by *Alternaria alternaria* (Fr. : Fr.) Keissl. f. *fragaria* Dingey, occurs in Europe, New Zealand, Korea and Japan (Cho and Moon, 1980). Lesions appear on upper leaves as circular or irregular patches (2–5 mm) that are brown with dark purplish margins. Leaf blotch, *Gnomonia comari* P. Karst., is a sporadic problem in Europe and North America, depending on cultivar susceptibility (Bolay, 1971). It produces purplish to brownish blotches on young leaves and large, light brown necrotic spots on older leaves. Purple leaf spot, *Mycosphaerella* spp., has symptoms very similar to leaf scorch and has been reported in south-eastern United States, the United Kingdom and India.

Fungal diseases of the fruit

Grey mould

Grey mould (ash mould) or botrytis fruit rot, *Botrytis cinerea* Pers.: Fr., is a problem wherever strawberries are grown and is probably the most serious fruit rot in strawberries. Infected young blossoms in a cluster often show a browing and drying (blasting) that travels gradually down the pedicel (Bristow *et al.*, 1986). Fruit infections appear as soft, light brown, rapidly enlarging areas on the fruit. Fruit eventually mummify, with a grey, dusty powder covering them (Plate 15). Spread is rapid through handling, and damaged, mature fruit can be especially susceptible to grey mould after picking. Healthy berries can become a rotted mass within 48 h. Disease development is favoured by wet conditions and warm temperatures (Bulger *et al.*, 1987; Wilcox and Seem, 1994; Sosa-Alvarez *et al.*, 1995). During the blossom blight stage, the fungus initially colonizes senescing flowers during the bloom period. It enters green fruit through the flowers and remains inactive, until the fruit matures. Fruit infection is most severe in the densely shaded areas of the plant where berries are touching bare soil or other berries.

Leather rot

Leather rot of fruit, caused by *Phytophthora cactorum* (Lebert & Cohn) J. Schröt. is an important strawberry disease in the USA, Europe and parts of Asia. It is usually not a serious problem except in periods of excessive rain during fruiting in temperate regions. The pathogen can infect fruit at any stage of development. In green fruit, diseased areas are dark brown or natural green outlined with a brown margin, and the whole berry eventually becomes brown, with a rough texture and is leathery in appearance (Plate 16). Infected ripe fruit show less dramatic symptoms, but can be identified by their duller colour, a bitter taste, and a marked darkening of the xylem connections to seeds. Eventually the mature fruits will become tough and leathery, and sometimes a white mouldy growth appears on them. A distinct pathotype of *P. cactorum* also causes a crown rot and wilt of strawberry (see below).

The fungus overwinters as oospores within mummified fruit on the soil. In the spring these oospores germinate in the presence of free water and produce sporangia with zoospores. These zoospores swim to fruit surfaces, germinate and infect the fruit. In the later stages of infection, sporangia are produced on the surface of infected fruit. During the growing season, both sporangia and zoospores are spread by splashing or wind blown water (Madden and Ellis, 1990; Madden *et al.*, 1991).

Anthracnose

Anthracnose fruit rot (black spot) is caused by several related pathogens including, *Colletotrichum fragariae* A.N. Brooks, *C. acutatum* J.H. Simmonds and *C. gloeosporioides* (Penz.) Penz. & Sacc. Anthracnose attacks foliage, runners, crowns and fruit (Smith and Black, 1990; Gunnell and Gubler, 1992; Maas, 1998). *C. fragariae* is the most common cause of crown rot in the southeastern USA, whereas *C. acutatum* is the principal fruit rot in this region (Howard *et al.*, 1992). In Europe, *C. acutatum* is the primary pathogen, with *C. fragariae* apparently absent (Denoyes and Baudry, 1995). Affected stems are sometimes girdled by lesions that contain salmon-coloured massses of spores. Infected crowns turn reddish brown, and the entire plant may wilt and die. Symptoms on all stages of fruit appear first as white, water-soaked lesions up to 3 mm in diameter and these become brown (Plate 17) and ultimately encompass the whole fruit. The lesions also contain salmon-coloured spore masses. Fruit eventually dry to hard, black shrivelled mummies. Spores are spread from infected plants to healthy ones by splashing water.

Other fruit rot diseases

There are a number of other fruit rot diseases of occasional worldwide economic importance (Maas, 1984). Rhizoctonia fruit rot (hard rot), *Rhizoctonia solani* Kuehn, is an important pathogen of many worldwide crops and is most severe where fruit are allowed to touch the soil. Symptoms include light brown fruit that soften and collapse, with their juices leaking out (Aerts, 1978). Rhizopus rot (leak), *Rhizopus* spp., which can occur worldwide, has caused severe problems in the preserve industry of the United Kingdom (Maas, 1998; Harris and Dennis, 1980). Symptoms include slightly discoloured fruits that gradually turn light brown (Dennis and Mountford, 1975). The fruit eventually collapse, become covered with a dense, fluffy mycelium, and their juices leak out. The importance of rhizopus rot is greatly diminished by rapid cooling after harvest. Mucor fruit rot, *Mucor* spp., can be important in the USA and UK. The pathogen produces pectolytic enzymes that degrade the fruit structure, and can be a major problem in the manufacture of preserves (Smith *et al.*, 1979; Harris and Dennis, 1980). Tan-brown rot, *Discohainesia oenotherae* (Cooke & Ellis) Nannf., is a major problem in warm, humid strawberry regions (Sutton and Gibson, 1977). It begins as small, tan slightly sunken spots that steadily enlarge; the centres of the spots go deep into the fruit, and the fruit eventually becomes dry and spongy as the mycelium replaces fruit tissue. Outbreaks of stem end rot, *Gnomonia comari* P. Karst., have occasionally occurred in North America and Europe, resulting in a brown discoloration in fruit near the calyx end (van der Scheer, 1981). Phomopsis (Dendrophoma) soft rot , *Phomopsis obscurans* (Ellis & Everh.) Sutton, is a very important fruit rot in Florida (Howard and Albregts, 1973). It is commonly associated with stem-end rot and appears initially as light-pink, water-soaked lesions on fruit that eventually turn brown with crusty centres. Septoria hard rot, *Septoria fragariae* (Lib.) Desm., is rare in North America, but is common in Europe and Australia. Hard, brown, sunken areas develop on immature fruits that eventually become lesions filled with clusters of seeds.

Control of leaf and fruit diseases

Fungicides can be used to control most of the common leaf diseases (Wilhelm and Nelson, 1990; Strand, 1994; Funt *et al.*, 1997), but as with insect pests, the key to good control of most leaf diseases is good sanitation. Disease spread is minimized to a large extent by maintaining good air flow through proper plant spacing and renovation, including removal of old leaves after harvest. Minimizing the amount of free water is critical, as species such as leaf spot need at least 12 h of continuous free moisture for infection to occur. Nursery fields have been shown to be an important source of some leaf diseases in fruiting fields, making good control in plant production fields critical (Gubler *et al.*, 1995). Biological control has not been developed for most leaf diseases, although microbial isolates have been identified with potential to control leaf scorch (Zheng, 1992; Sutton, 1994). Resistant varieties exist for most of the leaf pathogens.

Anthracnose of strawberry runners and petioles has been particularly difficult to control as there are no highly efficacious fungicides (Gubler *et al.*, 1995). Minimizing free water at all stages helps, and methyl bromide/chloropicrin fumigation of soil destroys soil-borne conidia. Conidia which contaminate nursery plants can also be effectively controlled by dipping in hot water (49°C) followed by rapid cooling at 12°C. The most effective method for controlling anthracnose is to maintain clean plant programmes. Clones resistant to anthracnose have been developed for breeding purposes (Smith and Gupton, 1993) and a number of resistant cultivars have been identified (Denoyes-Rothan and Guérin, 1996; Denoyes-Rothan, 1997).

A number of fungicides are also available for the control of fruit rots (Wilhelm and Nelson, 1990; Strand, 1994; Funt *et al.*, 1997). Although these are frequently used as prophylactics, several predictive models have been developed to target fungicide sprays based on disease cycles and environmental conditions (Reynolds *et al.*, 1987; Wilson *et al.*, 1990; Madden *et al.*, 1991, 1993). For example, most of the primary infection for botrytis fruit rot comes from leaf residue in the row and most infection occurs during bloom (Braun and Sutton, 1987, 1988). Therefore, the most efficient control strategies incorporate fungicide sprays primarily at bloom, good drainage and the removal or destruction of foliar residues (Bulger *et al.*, 1987; Sutton, 1990; Wilcox and Seem, 1994; Ellis *et al.*, 1998).

Culture can play an important role in rot control. Excessive use of nitrogen in the spring has been associated with increased levels of botrytis fruit rot. Foliar applications of calcium delays postharvest development of grey mould (Cheour *et al.*, 1990). Straw mulch has been shown to be very important in the control of leather rot, grey mould and anthracnose, both by directly protecting the fruit from inoculum and limiting splash dispersal of the fungal pathogens (Reynolds *et al.*, 1989; Madden and Ellis, 1990). The plastic mulches used in annual production systems actually increase the incidence of anthracnose and splash dispersal.

Several biological control methods for fruit pathogens show potential in strawberry, although they have not been widely utilized. Both *Trichoderma harzianum* and *Gliocladium roseum* have been shown to be effective against grey mould (Sutton and Peng, 1993; Sutton, 1994; Bélanger *et al.*, 1995). Isolates of *Bacillus pumilus*, *Gliocladium roseum*, *Trichoderma viride*, *Penicillium* sp. and *Pseudomonas fluorescens* appear effective as antagonists of *B. cinerea* (Sutton and Peng, 1993; Swadling and Jefferies, 1996). Methods have been developed to bee-vectored *Gliocladium roseum* for botrytis rot control (Peng *et al.*, 1992). Genotypes have been identified with some resistance to anthracnose and botrytis fruit rots (Ellis, 1995; Smith and Gupton, 1993), but the resistance is not strong enough to preclude the use of other control strategies (Ellis, 1995).

Fungal diseases of the roots and crowns

Several of the fungi that cause serious fruit rots, also infect crowns. Among the most damaging are *Colletotrichum fragariae* and *C. acutatum*, whose crown symptoms have already been described in the section on anthracnose. In addition, the leather rot organism, *Phytophthora cactorum*, causes a serious crown disease in Europe called crown rot or vascular collapse. Young leaves wilt suddenly, turn brown and the whole plant dies within a few days. Crowns break easily at their tops, leaving most of the crown behind. The crowns are brown with disintegrated vascular tissue. Infection occurs primarily in the runnering bed or after transplanting (Pettitt and Pegg, 1994). The source of inoculum is oospores in the soil or in infected plants. Zoospores infect plants through wounds. Plants appear most susceptible when temperatures are high in early to mid-summer, soil drainage is poor, and plants are stressed. Cultivars vary greatly in susceptibility to crown rot.

Red stele root rot

Red stele root rot (red core), *Phytophthora fragariae* Hickman, is a worldwide problem in strawberries and has caused significant recent losses in northeastern North America, France, Germany, Sweden and Switzerland (Maas, 1998). It is particularly troublesome in heavy clay soils that are saturated with water during the cool seasons. Optimal disease development occurs at 13 – 15°C (Duncan and Kennedy, 1995). Infected plants are stunted, produce few runners and lose their normal shiny green lustre (Plate 18). Younger leaves can have a metallic bluish-green colour, whereas older leaves can turn yellow or red. During hot, dry periods, diseased plants rapidly wilt and die. They produce few roots and the stele of the roots is pink to brick red (Fig. 8.5). The initial infection occurs mainly through the distribution of infected plants or infested soil. Oospores of the fungus produce large quantities

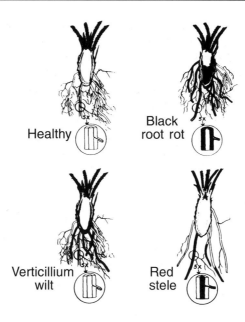

Fig. 8.5. How different types of rots affect roots (drawings from Perry and Ramsdell, 1983, with permission).

of zoospores that are mobile and swim about when soil moisture is high. The zoospores attack the tips of developing roots and move to the vascular system which they destroy. Once established, the fungus can survive in the soil for more than a decade.

Verticillium wilt

Verticillium wilt, *Verticillium albo-atrum* Reinke & Berth. and *V. dahlia* Kleb., is a disease of hundreds of crop plants, including strawberry. The fungus remains alive in a field for as long as 25 years. On infected plants, leaves wilt rapidly, dry and become reddish-yellow or dark brown. Few leaves and runners are formed. New roots are dwarfed with blackened tips (Fig. 8.5). In new plantings the symptoms begin to appear as runners are formed, whereas in established plantings symptoms are most prevalent just before picking. The fungus over-winters in soil or plant debris as dormant mycelium or microsclerotia. Under favourable conditions these germinate and produce hyphae which penetrate root hairs. Once inside the root, the fungus destroys the vascular system.

Black root rot

Black root rot, is caused by a complex of organisms including *Pythium* and *Rhizoctonia* species (Fig. 8.5). Various fungi, nematodes and abiotic factors have been implicated in the disease (LaMondia and Martin, 1989; Wing *et al.*,

1995). *Pythium* spp. by themselves cause a disease called pythium root rot or stunt (Watanabe *et al.*, 1977; Maas, 1998). Roots affected by black root rot have one or more of the following symptoms (Funt *et al.*, 1997): (i) much smaller root systems than normal; (ii) main roots are spotted with dark patches or zones; (iii) most feeder roots are lacking; (iv) all or part of the main root is dead and blackened throughout. Plants with black root rot are less vigorous and produce few runners. Often, black root rot is associated with nematodes, winter injury, fertilizer burn, soil compaction, herbicide damage, drought, excess salt, water logging or improper soil pH (Funt *et al.*, 1997). Scattered plants can be affected or localized regions within fields.

Other crown and root fungi

Several other root and crown fungi are of occasional worldwide importance (Maas, 1998). In *Rhizoctonia solani* root rot, plants suddenly collapse during early fruiting. The undersides of leaves become purplish and numerous side crowns are formed around dead ones (Van Adrichem and Bosher, 1962). Sclerotina crown rot, *Sclerotina sclerotiorum* (Lib.) de Bary, is primarily a crown rotting organism in the southern United States, but can also affect fruit (Sclerotina rot). Both crowns and fruit become covered with dense white mycelium containing sclerotina (Alcorn, 1966). Fusarium wilt (fusarium yellows), *Fusarium oxysporium* Schlecht. ex. Fr., is a problem primarily in Australia and Japan, although it is found sporadically elsewhere. Roots become rotted and leaves of infected plants wilt rapidly and crowns show a distinctive reddish brown discoloration and decay (Winks and Williams, 1965).

Nematodes

Besides being important vectors of strawberry diseases, nematodes can cause damage directly to strawberries. Most have a very wide species range. These organisms restrict root growth by feeding directly on roots and make them more susceptible to root rotting fungi. Affected plants are low in vigour, and show symptoms of mineral deficiencies and water stress. Their productivity is also greatly limited and in severe cases, the plants die.

A number of nematodes cause serious damage to strawberries (Howard *et al.*, 1985; Brown *et al.*, 1993). The dagger nematode, *Xiphinema* spp., is found all across temperate and subtropical regions. They cause sunken, reddish brown lesions on roots where they penetrate the epidermis and cortical cells. The leaf and stem nematode, *Ditylenchus dipsaci* (Kühn) Filipjev, is a cosmopolitan species that prefers cool, moist conditions, and when moisture is high it can be found in crowns, leaves and stems (Sturhan and Brzeski, 1991). These nematodes soften the tissue and cause stunting and deformation. They are found in Europe and the Pacific Northwest. The northern root-knot nematode *Meloidogyne hapla* Chitwood are found primarily in cooler production regions and form galls on fine roots where females feed and lay eggs (Edwards *et al.*, 1985). The root lesion nematode, *Pratylenchus* spp., is a worldwide pest that causes varying symptoms from brown lesions on roots to a black necrosis of the entire root system (Kurppa and Vrain, 1989). The spring dwarf nematode, *Aphelenchoides* spp., is a problem in northern Europe, the northern Atlantic and northwestern states of the USA. They live as ectoparasites on young buds and leaflets in the crown and leaf axils, causing stunting and deformation. The summer dwarf, *Aphelenchoides besseyi* Christie, and sting nematodes, *Belonolaimus longicaudatus* Rau, are primarily problems in Florida. The summer dwarf nematode causes stunting and deformation of new leaves with twisting, cupping and crinkling (Smart and Nguyen, 1991). Sting nematodes cause marginal browning in developing leaves and eventual drying of the leaf tissues adjacent to midribs (McCullock, 1978).

Control of root diseases and nematodes

Preplant fumigation with methyl bromide has proven to be the most effective way to control root rot fungi (Himelrick and Dozier, 1991), although this compound is set to be phased out by the 133 co-signers of the Montreal Protocol of the United Nations. Numerous other compounds have been evaluated as alternatives including 1,3-dichloropropene, metam sodium, chloropicrin and dazomet, but all are either less effective than methyl bromide or much more expensive (Anonymous, 1993; Noling and Becker, 1994; Stephens, 1996; Maas and Galletta, 1997). Solarization can prove useful in eradicating soil pests, but only works well in areas with long periods of high daily temperatures (Himelrick and Dozier, 1991; Gamliel and Stapleton, 1993; Hartz *et al.*, 1993).

Many of the root diseases can persist in the soil long after the initial infection, so it is critical that the soils are well drained to discourage fungal establishment and spread, and fields are started with clean planting stock. The factors most highly correlated with severe black root rot infestations in New York were lack of crop rotation, compacted or fine-textured soils, high rates of the herbicide terbacil and flat vs. raised bed culture (Wing *et al.*, 1995).

Several biological strategies have been tested in the control of soil-borne diseases. Attempts have been made to protect strawberries from *Fusarium oxysporum* f. sp. *fragariae* by previous inoculation with non-pathogenic strains of the same species (Tezuka and Makino, 1991) or soil amendments with a chitin-degrading yeast *Streptomyces anulatus* (Toyoda *et al.*, 1994). *Trichoderma* spp. have been used to control *Fusarium* wilt (Moon *et al.*, 1995) and post-transplant summer death (D'Ercole *et al.*, 1988). The biocontrol fungus *Gliocladium virens* (= *Trichoderma virens*) has potential for the control of *Phytophthora cactorum* (Wilcox, 1995). Root colonization by mycorrhizal fungus confers some resistance to *Phytophtora fragariae* (Norman *et al.*, 1996).

In some cases, rotations incorporating other crops and fallow has helped to reduce populations of the shorter-lived pathogens. Methods have been developed to predict disease risk for verticillium wilt (Harris and Yang, 1996) and red stele (Duncan, 1980). Probably the most important biological control strategy has been the development of cultivars with resistance to the diseases, particularly *P. fragariae* var. *fragariae* (Milholland and Daykin, 1993; Van de Weg *et al.*, 1993; see also Chapter 4). The development of sensitive detection techniques for pathogens in plants will also aid in reducing spread (Amouzou-Alladaye *et al.*, 1988; Mohan *et al.*, 1989).

Like soil-borne diseases, nematodes are most effectively controlled by fumigation (Strand, 1994). Other possibilities include non-fumigant carbamate and organophosphate nematicides (Maas, 1998), and the application of compounds produced by bacteria such as ivermectins (Potter, 1995), neurotransmitters (Nordmeyer, 1992), biological control organisms including *Pasteuria* bacteria and *Paecilomyces* fungi (Stirling, 1991), and soil amendments such as organic materials and chitin (Rodriguez-Kabana *et al.*, 1987; Vrain, 1990; Gamliel and Stapelton, 1993). Isolates of *Streptomyces* sp. show promise in controlling *Pratylenchus penetrans* (Dicklow *et al.*, 1993). Physical methods of control have been examined including sterilization utilizing hot water or steam, microwaves and solarization (Katan, 1981; Potter, 1995). In addition, resistance has been identified in both cultivars and wild species of strawberries (Potter and Dale, 1994).

Virus and virus-like diseases of strawberry

Several aphid-transmitted viruses are economically important (Table 8.4), but the severity of their symptoms is often cultivar specific and in most cases their effects are much more important in conjunction with other viruses than alone (Maas, 1998).

Strawberry mottle virus (SMV) is a worldwide problem caused by an unidentified virus (Leone *et al.*, 1992) that may be represented by a number of different strains and species. In susceptible cultivars, symptoms include small, chlorotic spots that are scattered across leaves, faintly to finely cleared veins, mild mottling, and minor to severe leaf distortion.

Strawberry crinkle virus (SCV) is a rhabdovirus that is a particularly important problem in the west coast of North America, but is also prevalent in Europe, New Zealand, Australia, South Africa and Japan. It represents several mild to severe strains, and is most damaging in combination with the pallidosis agent (see below) or the other aphid-borne viruses. The most characteristic symptoms are chlorotic spotting and distorted, crinkled leaves (Yoshikawa *et al.*, 1986; Hunter *et al.*, 1990).

Strawberry mild yellow-edge virus (SMYED) is another worldwide problem that has few diagnostic symptoms in the field, but in conjunction

Table 8.4. Known field vectors of virus and phytoplasm diseases in strawberry (adapted from Maas, 1998).

Disease	Vector	
Aster yellows	Leaf hoppers	*Macrosteles fascifrons*
		Colladonus geminatus
Arabis mosaic	Nematode	*Xiphinema diversicaudatum*
Crinkle	Aphid	*Chaetosiphon fragaefolii*
		C. jacobi
Green petal	Leaf hoppers	*Aphrodes bicincta*
		M. fascifrons
		Euscelis lineolata
		E. plebejus
Latent C	Aphid	*C. fragaefolii*
		C. thomasi
Latent ringspot	Nematode	*Xiphinema diversicaudatum*
Mild yellow edge	Aphid	*Chaetosiphon* sp.
Mottle	Aphid	*Chaetosiphon* sp.
		Aphis gossypii
Raspberry ringspot	Nematode	*Longidorus elongatus*
Tomato black ring	Nematode	*Longidorus elongatus*
Veinbanding Caulimovirus	Aphid	*Chaetosiphon* sp.
Tomato ringspot	Nematode	*Xiphinema americanum*

with other viruses result in large reductions in plant vigour, runner production and yield. Mild yellow-edge virus plus crinkle and/or mottle virus act synergistically to form the yellows complex, identified by marginal chlorosis, cupping of young leaves and reddening of older leaves. SMYED has traditionally been considered to be due to a luteovirus, but a potexvirus has been implicated recently in symptom expression in *F. vesca* (Jelkmann *et al.*, 1990).

Strawberry latent C virus (SLCV) is a rhabdovirus that is a problem in the eastern USA, Nova Scotia and Japan (Yoshikawa and Inouye, 1988; Maas, 1998). It does not produce consistent symptoms but can be degenerative. Its effects are additive to the other aphid transmitted viruses. Strawberry vein banding caulimovirus (SVBV) occurs occasionally in strawberry fields all across the world. Most of the cultivars now grown are symptomless, although older ones showed reduced runner production, yield and fruit quality. SVBV is most important when it is in mixed infections with one of the other strawberry viruses. It interacts more strongly with SVBV, SCV or SLCV than SMV, SMYEV or the pallidosis agent (Maas, 1998).

A number of strawberry viruses are nepoviruses that are transmitted by nematodes (Table 8.4) (Converse, 1987; Maas, 1998). These viruses can be lethal, but travel very slowly in the field and are generally of minor consequence unless planted into highly infected fields or are infected in the nursery.

Raspberry ringspot virus (RpRSV) is a serious disease in the UK and the former USSR. Symptoms include chlorotic rings and spots, general chlorosis and necrotic spots that vary in intensity, depending on the cultivar.

Tomato black ring virus (TBRV) produces similar symptoms and is also a serious disease in the UK and the former Soviet block.

Arabis mosaic virus (ArMV) is important in western Europe and causes mild to severe chlorotic leaf blotches, mild to severe stunting and moderate to severe leaf distortion in some cultivars.

Strawberry latent ringspot virus (SLRSV) is important in western Europe and the Slovic countries and is identified by dwarfing and yellow leaves. Tomato ringspot virus (ToRSV) is the only nepovirus of strawberries in North America, where it is a serious disease along the Pacific coast and in the Great Lakes region. There is a wide variety of symptoms found across cultivars from none to dwarfing, reduced runner numbers, mottling and leaf death (Maas, 1998).

Two other viruses with unknown vectors can cause serious diseases of strawberry. Strawberry pallidosis is a serious disease in the USA, eastern Canada and Australia. A specific vector has not been clearly determined, but the disease spreads in a pattern consistent with leaf hopper movement (Converse, 1987) and it has been experimentally transferred by *Coelidia olitoria* (Maas, 1998). A specific causative organism has also not been identified, but dsRNA have been found in a number of pallidosis isolates (Yoshikawa and Converse, 1990). It is latent in commercial cultivars, with the only external symptoms being a lack of vigour, marginal chlorosis of young leaves and premature reddening of older leaves. Tobacco streak iarvirus (TSI) occurs in the US, Israel and Australia. It is generally symptomless, but can cause substantial yield reductions (Maas, 1998). TSV is transmitted by thrips in annual crops, but there is no direct evidence of their involvement in strawberries.

Several diseases of strawberry are caused by phytoplasma and are transmitted by leaf hoppers (Table 8.4). What has long been described as Aster yellows is actually a descriptor for several diseases caused by a group of phytoplasma in taxonomic subgroup 16SrI (Lee *et al.*, 1992; Lee and Davis, 1993). They are a common problem in North America, Europe, Australasia, Japan and Russia. Older leaves turn red, lie flat on the ground, and the plant is usually dead within 2 months. Green petal (clover phyllody) is found throughout northeastern North America, Europe and Australia. It is caused by a phytoplasma in aster yellows 165 RNA taxonomic group I, subgroup C and may represent more than one phytoplasma. Its most obvious symptom is the production of virescent and adherent green petals, petals later turn red and fruits become 'buttoned' (Posnette and Chiykowski, 1987).

Control of virus and phytoplasma diseases

There are no effective means to control virus and phytoplasma diseases, except to plant virus-free stock and eliminate the insect and nematode vectors. Annual replanting can also prevent the buildup of viruses in plants over time. Some varieties also have resistance to common aphid vectors (Hancock *et al.*, 1990). Although the diagnosis of symptoms is often difficult in the field, numerous indicator genotypes of *F. vesca* have been developed for this purpose (Bringhurst and Voth, 1956; Frazier, 1974), and a growing number of ELISA and PCR tests are appearing (Stenger *et al.*, 1988; Kaden-Kreuzinger *et al.*, 1995; Martin, 1995; Schoen and Leone, 1995). The thermal treating of meristems is widely used to eliminate viruses from propagation stock (see Chapter 6).

CONCLUSIONS

Although many diseases and pests of strawberries have highly localized distributions, there are a number of problems that are worldwide. Aphids, spider mites and lygus bugs are the most widespread sucking and piercing insects. Flower thrips, root weevils and leafrollers are the most common chewing pests. Ramularia leaf spot is the most widespread leaf disease of strawberries, with leaf scorch, leaf blight and powdery mildew also being quite common. Grey mould, leather rot and anthracnose are the most widespread fruit rots. The fungal diseases red stele and verticillium wilt infest roots all across the world, together with the dagger and root lesion nematodes. Angular leaf spot and crown rot are more localized in their distribution, but are probably the most rapidly spreading diseases.

Effective chemical controls have been developed for most of these pests and diseases, but recent emphasis has been placed on minimizing these chemical inputs. Integrated crop management strategies are emerging that incorporate cultural, genetic and biological control methods in conjunction with regular monitoring of pest populations. Some of the most comprehensive ICM strategies have been developed for spider mites and lygus bugs, but effective means to minimize chemical use are also available for most diseases and pests. Maintaining good air flow, minimizing standing water, starting with clean planting stock and utilizing resistant varieties are key to the control of most fungal pathogens. Isolation, renovation and scouting are critical to reducing most insect pests and a wide array of biological control agents have also been developed that show high promise.

Unfortunately, we are still struggling with the problem of finding an alternative to methyl bromide fumigation as a means of controlling soil diseases and pests. It is expected that chemical alternatives will be found, but it seems that a stronger effort should be made to identify sources of genetic resistance. There may be a yield reduction associated with the incorporation of resistance genes, but the economic loss may be less than the cost of future methods of chemical control.

REFERENCES

Aalders, L.E. and Craig, D.L. (1968) General and specific combining ability in seven inbred strawberry lines. *Canadian Journal of Genetics and Cytolology* 10, 1–6.

Abdelrahman, M.H. (1984) Growth and productivity of strawberry cultivars at high temperatures. Ph.D. Diss., Kansas State University, Manhattan (Diss. Abst. Inter. B 46:17-B).

Abeles, F.B. and Takeda, F. (1990) Cellulase activity and ethylene in ripening strawberry and apple fruits. *Scientia Horticulturae* 42, 269–275.

Abers, J.E. and Wrolstad, R.E. (1979) Causative factors of color deterioration in strawberry preserves during processing. *Journal of Food Science* 44, 75–82.

Adams, J.N. and Ongley, M.H. (1973) The degredation of anthocyanins in canned strawberries. I. The effect of various processing parameters on the retention of pelargonidin-3-glucoside. *Journal of Food Chemistry* 8, 139–145.

Aerts, J. (1978) Bestrijding van Rhizoctonia solani Kühn in glasardbeien. *Mededelingen Faculteit Landbouwwetenschappen Rijksuniversiteit te Gent* 43, 997–1005.

Aharoni, Y. and Barkai-Golan, R. (1987) Pre-harvest fungicide sprays and polyvinyl wraps to control *Botrytis* rot and prolong the postharvest storage of strawberries. *Journal of the Horticultural Society* 62, 177–181.

Ahmadi, H. and Bringhurst, R.S. (1991) Genetics of sex expression in *Fragaria* species. *American Journal of Botany* 78, 504–514.

Ahmadi, H. and Bringhurst, R.S. (1992) Breeding strawberries at the decaploid level. *Journal of the American Society for Horticultural Science* 117, 856–862.

Ahmadi, H., Bringhurst, R.S. and Voth, V. (1991) Modes of inheritance of photoperiodism in *Fragaria*. *Journal of the American Society for Horticultural Science* 115, 146–152.

Ahokas, H. (1995) *Fragaria virginiana* and *F. vesca* on the southern coast of Finland. *Annales. Botanica Fennici* 32, 29–33.

Albregts, E.E. and Howard, C.M. (1972) Comparison of polyethylene and polyethylene-coated bio-degradeable paper mulches on strawberry. *HortScience* 7, 568–569.

Albregts, E.E. and Howard, C.M. (1986) Effect of runner removal on strawberry fruiting response. *HortScience* 21, 97–98.

Alcorn, J.W.S. (1966) *Sclerotinia sclerotiorum* on strawberry. *Horticultural Research* 6, 128.

Alpert, P. (1991) Nitrogen sharing among ramets increases clonal growth in *Fragaria chiloensis*. *Ecology* 72, 69–80.

Alpert, P. (1996) Nutrient sharing in natural clonal fragments of *Fragaria chiloensis*. *Journal of Ecology* 84, 395–406.

Alpert, P. and Mooney, H.A. (1986) Resource sharing among ramets in the clonal herb, *Fragaria chiloensis*. *Oecologia* 70, 227–233.

Amouzou-Alladaye, E., Dunez, J. and Clerjeau, M. (1988) Immunoenzymic detection of *Phytophthora fragariae* in infected strawberry plants. *Phytopathology* 78, 1022–1026.

Anderson, H.M. (1982) The cold-storage of strawberry runners – a review. *Crop Research* 22, 93–104.

Anderson, J.A. and Whitworth, J. (1993) Supercooling strawberries plants inoculated with ice-nucleation-active bacteria and tested with Frostguard. *HortScience* 28, 828–830.

Anon. (1993) *Alternatives to Methyl Bromide: Assessment of Research, Needs, and Priorities.* United States Department of Agriculture Global Change Program Office, Arlington, Virginia.

Anon. (1995) *US Strawberry Industry.* Economic Research Service, USDA, Washington, DC.

Anstey, T.H. and Wilcox, A.N. (1950) The breeding value of selected inbred clones of strawberries with respect to their vitamin C content. *Scientific Agriculture* 30, 367–374.

Antonelli, A.L., Shanks, C.H. Jr and Fisher, G.C. (1991) *Small Fruit Pests: Biology, Diagnosis and Management.* Washington State University, Cooperative Extension, Bulletin EB1388.

Antonelli, A.L., Mayer, D.F., Burgett, D.M. and Sjulin, T. (1988) Pollinating insects and strawberry yields in the Pacific Northwest. *American Bee Journal* 128, 618–620.

Antoszewski, R. and Dzieciol, U. (1973) Translocation and accumulation of ^{14}C-photosynthates in the strawberry plant. *Horticultural Research* 13, 75–81.

Archbold, D.D. (1988) Abscisic acid facilitates sucrose import by strawberry fruit explants and cortex discs *in vitro*. *HortScience* 23, 880–881.

Archbold, D.D. (1993) Foliar attributes contributing to drought stress tolerance in *Fragaria* species. *Acta Horticulturae* 348, 347–350.

Archbold, D.D. (1996) Drought resistance in the strawberry: what is the potential for improvement? In: Pritts, M.P., Chandler, C.K. and Crocker, T.E. (eds) *Proceedings of the IV North American Strawberry Conference*, University of Florida, Orlando, pp. 127–132.

Archbold, D.D. and Dennis, F.G. Jr (1984) Quantification of free ABA and conjugated IAA in strawberry achene and recepticle tissue during fruit development. *Journal of the American Society for Horticultural Science* 109, 330–335.

Archbold, D.D. and Zhang, B. (1991) Drought stress resistance in *Fragaria* species. In: Dale, A. and Luby, J.J. (eds) *The Strawberry into the 21st Century*. Timber Press, Portland, Oregon, pp. 138–144.

Archbold, D.D., Hamilton-Kemp, T.R., Barth, M.M. and Langlois, B.E. (1997a) Identifying natural volatile compounds which control grey mold (*Botrytis cinerea*) during postharvest storage of strawberry, blackberry, and grape. *Journal of Agricultural Food Chemistry* 45, 4032–4037.

Archbold, D.D., Hamilton-Kemp, T.R., Langlois, B.E. and Barth, M.M. (1997b) Natural volatile compounds control *Botrytis* on strawberry fruit. *Acta Horticulturae* 439, 923–930.

Arney, S.E. (1953a) Studies of growth and development in the genus *Fragaria*. I. Factors effecting the rate of leaf production in Royal Sovereign strawberry. *Journal of Horticultural Science* 28, 73–84.

Arney, S.E. (1953b) Studies of growth and development in the genus *Fragaria*. II. The initiation, growth and emergence of leaf primordia in *Fragaria*. *Annals of Botany* 17, 476–492.

Arney, S.E. (1954) Studies of growth and development in the genus *Fragaria*. III. The growth of leaves and roots. *Annals of Botany* 18, 349–365.

Arulsekar, S. (1979) Verticillium wilt resistance in the cultivated strawberries and preliminary studies on isozyme genetics in *Fragaria*. PhD thesis. University of California, Davis, California.

Arulsekar, S. and Bringhurst, R.S. (1981) Genetic model for the enzyme marker PGI in diploid California *Fragaria vesca*. *Journal of Heredity* 73, 117–120.

Arulsekar, S., Bringhurst, R.S. and Voth, V. (1981) Inheritance of PGI and LAP isozymes in octoploid cultivated strawberries. *Journal of the American Society for Horticultural Science* 106, 679–683.

Asker, S. (1970) An intergeneric *Fragaria* × *Potentilla* hybrid. *Hereditas* 64, 135–139.

Austin, M.E. (1991) Rowcovers for 'Sparkle' strawberries. *HortScience* 26, 603.

Austin, M.E., Shutak, V.G. and Cristopher, E.P. (1960) Color changes in harvested strawberry fruits. *Proceedings of the American Society for Horticultural Science* 75, 382–386.

Avigdori-Avidov, H. (1986) Strawberry. In: Monselise, S.P. (ed.) *CRC Handbook of Fruit Set and Development*. CRC Press, Boca Raton, Florida, pp. 419–449.

Backhaus, G.F. (1994) Biological control of *Otiorhynchus sulcatus* by use of enthomopathogenic nematodes of the genus *Heterorhabditis*. *Acta Horticulturae* 364, 131–142.

Baker, R.E. (1952) Inheritance of fruit characters in the strawberry: a study of several F_1 hybrid and inbred populations. *Journal of Heredity* 43, 9–14.

Bakker, J., Bridle, P. and Bellworthy, S.J. (1994) Strawberry juice color: a study of the quantitative and qualitative pigment composition of juices from 39 genotypes. *Journal of Science, Food and Agriculture* 64, 31–37.

Barrientos, F. and Bringhurst, R.S. (1973) A haploid of an octoploid strawberry cultivar. *HortScience* 8, 44.

Barritt, B.H. (1974) Single harvest yields of strawberries in relation to cultivar and time of harvest. *Journal of the American Society for Horticultural Science* 99, 6–8.

Barritt, B.H. (1976) Evaluation of strawberry parent clones for easy calyx removal. *Journal of the American Society for Horticultural Science* 91, 267–273.

Barritt, B.H. (1979) Breeding strawberries for fruit firmness. *Journal of the American Society for Horticultural Science* 104, 663–665.

Barritt, B.H. (1980) Resistance of strawberry to *Botrytis* fruit rot. *Journal of the American Society for Horticultural Science* 105, 160–164.

Barritt, B.H. and Daubeny, H.A. (1982) Inheritance of virus tolerance in strawberry. *Journal of the American Society for Horticultural Science* 107, 278–282.

Barritt, B. H. and Shanks, C.H. (1980) Breeding strawberries for resistance to aphids *Chaetosiphon fragaefolii* and *C. thomasi. HortScience* 15, 287–288.

Barritt, B.H. and Shanks, C.H. (1981) Parent selection in breeding strawberries resistant to two-spotted spider mites. *HortScience* 16, 323–324.

Barritt, B.H., Bringhurst, R.S. and Voth, V. (1982) Inheritance of early flowering in relation to breeding of day-neutral strawberries. *Journal of the American Society for Horticultural Science* 107, 733–736.

Bartual, R., López-Aranda, J., Marsal, J.I., López-Montero, R., Castell, R., Barceló, V. and Juarez, J. (1993) Strawberry breeding in Spain: present situation and perspectives. *Acta Horticulturae* 348, 56–60.

Bartual, R., López-Aranda, J., Marsal, J.I., Medina, J.J., López-Montero, R. and López-Medina, J. (1997) Calderona: a new public Spanish strawberry cultivar. *Acta Horticulturae* 439, 261–264.

Battey, N.H., Le Mière, P., Tehranifar, A., Cekic, C., Taylor, S., Shrives, K.J., Hadley, P., Greenland, A.J., Darby, J. and Wilkinson, M.J. (1998) Genetic and environmental control of flowering in strawberry. In: Cockshull, K.E., Gray, D., Seymore, G.B. and Thomas, B. (eds) *Genetic and Environmental Manipulation of Horticultural Crops.* CAB International, Wallingford, UK.

Bauer, A. (1993) Progress in breeding decaploid *Fragaria × vescana. Acta Horticulturae* 348, 60–63.

Baumann, T. E. and Daubeny, H.A. (1989). Evaluation of the waiting-bed cultural system for strawberry season extension in British Columbia. *Advances in Strawberry Production* 8, 55–57.

Bazzocchi, R. (1968) Osservazioni sulla resistenza della cultivar di fragola alle gelate primaverili. (Observations on the resistance of strawberry varieties to spring frosts.) *Rivista Ortoflorofruttic.* 52, 394–400.

Bedard, P.R., Hsu, C.S., Spangelo, L.P.S., Fejer, S.O. and Rouselle, G.L. (1971) Genetic, phenotypic, and environmental correlations among 28 fruit and plant characters in the cultivated strawberry. *Canadian Journal of Genetics and Cytology* 13, 470–479.

Beech, M. G., Crisp, C.M., Thomas, C.M.S. and Wickenden, M.F. (1986) Extending the strawberry season – cultural systems for waiting-bed plants and extended cropping varieties. *Report of the East Malling Research Station for 1985*, East Malling, UK, p. 132.

Beech, M.G., Crisp, C.M., Simpson, S.E. and Atkinson, D. (1988) The effect of in vitro cytokinin concentration on the fruiting and growth of conventionally propagated strawberry runner progeny. *Journal of Horticultural Science* 63, 77–81.

Bélanger, R.R., Durour, N., Caron, J. and Benhamou, N. (1995) Chronological events associated with antagonistic properties of *Trichoderma harzianum* against *Botrytis cinerea*: indirect evidence for sequential role of antibiosis and parasitism. *Biocontrol Science and Technology* 5, 41–53.

Bell, J.A. and Simpson, D.W. (1994) The use of isoenzyme polymorphisms as an aid for cultivar identification in strawberry. *Euphytica* 77, 113–117.

Bell, J.A., Simpson, D.W. and Harris, D.C. (1997) Development of a method for screening strawberry germplasm resistant to *Phytophthora cactorum. Acta Horticulturae* 439, 175–179.

Berger, K.C. (1962) Micronutrient deficiencies in the US. *Agricultural Food Chemistry* 10, 178–181.

Binghurst, R.S. and Khan, D.A. (1963) Natural pentaploid *F. chiloensis – F. vesca* hybrids in coastal California and their significance in polyploid *Fragaria* evolution. *American Journal of Botany* 50, 658–661.

Bish, E.B., Cantliffe, D.J., Hockmuth, G.J. and Chandler, C.K. (1997) Development of containerized strawberry transplants for Florida's winter production system. *Acta Horticulturae* 439, 403–406.

Bite, A., Laugale, V. and Jurevica, Dz. (1997) Strawberry culture in Latvia. *Acta Horticulturae* 439, 403–406.

Bolay, A. (1971) Contribution à la connaissance de *Gnomonia comani* Karstan. Etude taxonomique, phytopathologique et recherches sur sa croissance *in vitro. Bericht den Schweizerischen Botanischen Gesellschaft* 81, 398–482.

Bors, B and Sullivan, J.A. (1998) Interspecific crossability of nine diploid *Fragaria* species. *HortScience* 32, 439.

Bostanian, N.J. (1994) The tarnished plant bug and strawberry production. *Agri-Food Canada Research Branch Technical Bulletin* 1994–1E.

Bostanian, N.J. and Mailloux, G. (1994) Threshold levels and sequential sampling plans for tarnished plant bug in strawberries. In: Bostanian, N.J., Wilson, L.T. and Dennehy, T.J. (eds) *Monitoring and Integrated Management of Arthropod Pests of Small Fruit Crops.* Intercept, Andover, UK, pp. 81–94.

Bostanian, N.J., Wilson, L.T. and Dennehy, T.J. (1994) *Monitoring and Insect Pest Management of Arthropod Pests of Small Fruit Crops.* Intercept, Andover, UK.

Boxus, P. (1974) The production of strawberry plants by *in vitro* micropropagation. *Journal of Horticultural Science* 49, 209–210.

Boxus, P., Quoirin, M. and Laine, J.M. (1977) Large scale propagation of strawberry plants from tissue culture. In: Reinert, J. and Bajaj, Y.P.S. (eds) *Applied and Fundamental Aspects of Plant, Cell, Tissue and Organ Culture.* Springer-Verlag, New York, pp. 130–143.

Boxus, P., Damiano, C. and Brasseur, E. (1984) Strawberry. In: Ammirato, P.V., Evans, D.A., Sharp, W.R. and Yamada, Y. (eds) *Handbook of Plant Cell Culture*, Vol. 3. Macmillan, New York, pp. 518–548.

Boyce, B.R. and Heleba, D. A. (1994) Comparison of several mulching systems for winter injury protection and yield of 'Midway' strawberry plants. *Advances in Strawberry Research* 13, 32–35.

Boyce, B.R. and Smith, C.R. (1967) Low temperature crown injury of dormant 'Catskill' strawberries. *Proceedings of the American Society for Horticultural Science* 91, 261-266.

Boyce, B.R. and Matlock, D.L. (1966) Strawberry nutrition. In: Childers, N.F. (ed.) *Nutrition of Fruit Crops.* Horticultural Publications, Rutgers University, New Brunswick, New Jersey, pp. 518–548.

Braun, P.G. and Sutton, J.C. (1986) Management of strawberry grey mold with fungicides targeted against inoculum in crop residues. *Brighton Crop Protection Conference on Pests and Diseases.* British Crop Protection Council, Surrey, pp. 915–921.

Braun, P.G. and Sutton, J.C. (1987) Inoculum sources of *Botrytis cinerea* in fruit rot of strawberry in Ontario. *Canadian Journal of Plant Pathology* 9, 1–5.

Braun, P.G. and Sutton, J.C. (1988) Infection cycles and population dynamics of *Botrytis cinerea* in strawberry leaves. *Canadian Journal Plant Pathology* 10, 133–141.

Breakey, E.P. and Dailey, E. F. (1956) A method for identifying cyclamen mite damage on Northwest variety strawberry plants. *State College of Washington Extension Service Circular* 261.

Brecht, J.K., Sargent, S.A., Bartz, J.A., Chau, K.V. and Emond, J.P. (1992) Irradiation plus modified atmospheric storage of strawberries. *Proceedings of the Florida State Horticulture Society* 105, 97–100.

Bringhurst, R.S. (1990) Cytogenetics and evolution in American *Fragaria*. *HortScience* 25, 879–881.

Bringhurst, R.S. and Gill, T. (1970) Origin of *Fragaria* polyploids. II. Unreduced and double-unreduced gametes. *American Journal of Botany* 57, 969–976.

Bringhurst, R.S. and Khan, D.A. (1963) Natural pentaploid *F. chiloensis* – *F. vesca* hybrids in coastal California and their significance in polyploid *Fragaria* evolution. *American Journal of Botany* 50, 658–661.

Bringhurst, R.S. and Senanayake, Y.D.A. (1966) The evolutionary significance of natural *Fragaria chiloensis* × *F. vesca* hybrids resulting from unreduced gametes. *American Journal of Botany* 53, 1000–1006.

Bringhurst, R.S. and Voth, V. (1956) Strawberry virus transmission by grafting excised leaves. *Plant Disease Reporter* 40, 596–600.

Bringhurst, R. S. and Voth, V. (1957) Effect of stratification on strawberry seed germination. *Proceedings of the American Society for Horticultural Science* 70, 144–149.

Bringhurst, R.S. and Voth, V. (1960) Larger strawberries through plant breeding. *California Agriculture* 14, 8.

Bringhurst, R.S. and Voth, V. (1975) Breeding and exploitation of day-neutral strawberries in subtropical environments. *HortScience* 10, 329.

Bringhurst, R.S. and Voth, V. (1978) Origin and evolutionary potentiality of the day-neutral trait in octoploid *Fragaria*. *Genetics* 90, 510.

Bringhurst, R.S. and Voth, V. (1984) Breeding octoploid strawberries. *Iowa State University Journal of Research* 58, 371–381.

Bringhurst, R.S., Voth, V. and Van Hook, D. (1960) Relationship of root starch content and chilling history to performance of California strawberries. *Proceedings of the American Society for Horticultural Science* 75, 373–381.

Bringhurst, R.S., Wilhelm, S. and Voth, V. (1961) Pathogen variability and breeding verticillium wilt resistant strawberries. *Phytopathology* 51, 786–794.

Bringhurst, R.S., Wilhelm, S. and Voth, V. (1966) Verticillium wilt resistance in natural populations of *Fragaria chiloensis* in California. *Phytopathology* 51, 219–222.

Bringhurst, R.S., Hansche, P.E. and Voth, V. (1967) Inheritance of verticillium wilt resistance and the correlation of resistance with performance traits of the straw–berry. *Proceedings of the American Society for Horticultural Science* 92, 369–375.

Bringhurst, R.S., Arulsekar, S., Hancock, J.F. and Voth, V. (1981) Electrophoretic characterization of strawberry cultivars. *Journal of the American Society for Horticultural Science* 106: 684–687.

Bringhurst, R.S., Voth, V and Shaw, D. (1990). University of California strawberry breeding. *HortScience* 25, 834.

Bristow, P.R., McNichol, R.J. and Williamson, B. (1986) Infection of strawberry flowers by *Botrytis cinerea* and its relevance to grey mold development. *Annals of Applied Biology* 109, 545–554.

Brooks, R.M. and Olms, H.P. (1997) *Register of Fruit and Nut Varieties*, 3rd edn. ASHS Press, Alexandria, Virginia.

Broome, O.C. and Goff, L.M. (1987) The production, testing and certification of virus-tested strawberry stock. *Advances in Strawberry Production* 6, 3–5.

Brouwer, R. (1963) Some physiological aspects of the influence of growth factors in the root medium on growth and dry matter production. *Medelel. Inst. Biol. Scheik. Onderz. Landbouwgewassen* 1, 1–30.

Brown, D.J.F., Dalmasso, A. and Trudgill, D.L. (1993) Nematode pests of soft fruits and vines. In: Evans, K., Trudgill, D.L. and Webster, J.M. (eds) *Plant Parasitic Nematodes in Temperate Agriculture*. CAB International, Wallingford, UK.

Brown, G.R. and Moore, J.N. (1975) Inheritance of fruit detachment in strawberry. *Journal of the American Society for Horticultural Science* 100, 569–572.

Brown, G.R., Moore, J.N. and Bailey, L.F. (1975) Evaluating genetic sources of fruit detachment traits in strawberry. *HortScience* 10, 120–121.

Brown, K.M., Geeson, J.D. and Dennis, C. (1984) The effects of harvest date and CO_2-enriched storage atmospheres on the storage and shelf-life of strawberries. *Journal of the Horticultural Society* 59, 197–204.

Brown, T. and Wareing, P.F. (1965) The genetical control of the everbearing habit and three other characters in varieties of *Fragaria vesca*. *Euphytica* 14, 97–112.

Bryant, L.R., Carstens, M.W. and Crandall, P.C. (1961) Strawberry plant storage. *Circular of the Washington Agricultural Experiment Station* 390, 1–5.

Buckley, B. and Moore, J.N. (1982) The effects of bed height, bed width, plant spacing, and runner removal on strawberry yield and fruit size. *Advances in Strawberry Production* 1, 4–5.

Bühler, J. (1922) *Schriften der Heiligen Hildegard von Bingen*. Insel Verlag, Leipzig.

Bulger, M.A., Ellis, M.A. and Madden, L.V. (1987) Influence of temperature and wetness duration on infection of strawberry flowers by *Botrytis cinerea* and disease incidence of fruit originating from infected flowers. *Phytopathology* 77, 1225–1230.

Bunday, L.G. and Bremner, J.M. (1974) Inhibition of nitrification in soils. *Soil Science Society of America* 37, 396–398.

Bunyard, E.A (1914) The history and development of the strawberry. *Journal of the Royal Horticultural Society* 39.

Burke, M.J., Gusta, L.V., Quamme, H.A., Weiser, C.J. and Li, P.H. (1976) Freezing and injury in plants. In: Briggs, W.R., Green, P.B. and Jones, R.L. (eds) *Annual Review of Plant Physiology*, Annual Reviews, Palo Alto, California, pp. 507–528.

Byrne, D. and Jelenkovic, G. (1976) Cytological diploidization in the cultivated octoploid strawberry *F. × ananassa*. *Canadian Journal of Genetics and Cytology* 18, 653–659.

Caldwell, J., Hancock, J.F. and Flore, J.A. (1990) Strawberry leaf photosynthetic acclimation to temperature. *HortScience* 95, 166.

Cameron, J.S. (1986) Factors influencing phenotypic variability in micropropagated strawberry (*Fragaria × ananassa*) cultivars. PhD Thesis. Michigan State University, East Lansing, Michigan.

Cameron, J.S. and Hancock, J.F. (1985) The field performance of strawberry nursery stock produced originally from runners or micropropagation. *Recent Advances in Strawberry Production* 4, 56–58.

Cameron, J.S. and Hancock, J.F. (1986) Enhanced vigor in vegetative progeny of micropropagated strawberry plants. *HortScience* 21, 1225–1226.

Cameron, J.S. and Hartley, C.A. (1990) Gas exchange characteristics of *Fragaria chiloensis* genotypes. *HortScience* 25, 327–329.

Cameron, J. S., Hancock, J.F. and Flore, J.A. (1989) The influence of micropropagation on yield components, dry matter partitioning and gas exchange characteristics of strawberry. *Scientia Horticulturae* 38, 61–67.

Cameron, J. S., Sjulin, T.M. , Shanks, C.H., and Muñoz, C.E. (1991) Collection of *Fragaria chiloensis* in central and southern Chile. In: Dale, A. and Luby, J. (eds) *The Strawberry into the 21st Century*. Timber Press, Portland, Oregon, pp. 108–110.

Cameron, J.S., Sjulin, T.M., Ballington, J.R., Shanks, C.H., Muñoz, C. and Lavín, A. (1993). Exploration, collection and evaluation of Chilean *Fragaria*: summary of 1990 and 1992 expeditions. *Acta Horticulturae* 348, 65–74.

Catling, P.M. (1995) *Fragaria multicipita*, reduced to the rank of forma. *Rhodora* 97, 245–254.

Catling, P.M. and Porebski, S. (1998) An ecoregional analysis of morphological variation in British Columbia coastal strawberries (*Fragaria*) for germplasm protection. *Canadian Journal of Botany* 78, 117–124.

Ceponis, M.J., Cappellini, R.A. and Lightner, G.W. (1987) Disorders in sweet cherry and strawberry shipments to the New York market, 1972–1984. *Plant Disease* 71, 472–475.

Chabot, B.F. (1978) Environmental influences on photosynthesis and growth in *Fragaria vesca*. *New Phytologist* 80, 87–98.

Chabot, B.F. and Chabot, J.F. (1977) Effects of light and temperature on leaf anatomy and photosynthesis in *Fragaria vesca*. *Oecologia* 26, 363–377.

Chabot, B.F., Jurik, T.M. and Chabot, J.F. (1979) Influence of instantaneous and integrated light-flux density on leaf anatomy and photosynthesis. *American Journal of Botany* 66, 940–945.

Chagnon, M., Gingras, J. and DeOliveira, D. (1993) Complementary aspects of strawberry pollination by honey and indigenous bees (Hymenoptera). *Journal of Ecological Entomology* 86, 416–420.

Chandler, C.K., Albregts, E.E., Howard, C.M. and Dale, A. (1989) Influence of propagation site on the fruiting of three strawberry clones in a Florida winter production system. *Proceedings of the Florida State Horticultural Society* 102, 310–312.

Chang, C.P. and Huang, S.C. (1995) Evaluation of effectiveness of releasing green lacewing, *Mallada basalis* (Walker) for the control of tetranichid mites in strawberry. *Plant Protection Bulletin* (Taiwan) 37, 41–58.

Cheng, G.W. and Breen, P.J. (1991) Activity of phenylalanine ammonia-lyase (PAL) and concentrations of anthocyanins and phenolics in developing strawberry fruit. *Journal of the American Society for Horticultural Science* 116, 865–869.

Cheour, F., Willemot, C., Arul, J., Makhlouf, J. Charest, P.M. and Gosselin, A. (1990) Foliar application of calcium chloride delays postharvest ripening of strawberry. *Journal of the American Society for Horticultural Science* 115, 789–792.

Cheour, F., Willemot, C., Arul, J., Makhlouf, J., Charest, P.M. and Desjardins, Y. (1991) Postharvest response of two strawberry cultivars to foliar application of CaCl$_2$. *HortScience* 26, 1186–1188.

Chercuitte, L., Sullivan, J.A., Desjardins, Y.D. and Bedard, R. (1991) Yield potential and vegetative growth of summer planted strawberry. *Journal of the American Society for Horticultural Science* 116, 930–936.

Chiykowski, L.N. and Craig, D.L. (1975) Reaction of strawberry cultivars to clover phyllody (green petal) agent transmitted by *Aphrodes bicincta. Canadian Plant Disease Survey* 55, 66–68.

Cho, J.T. and Moon, B.J. (1980) The occurrence of strawberry black leaf spot caused by *Alternaria alternata* (Fr.) Keissler in Korea. *Korean Journal of Plant Protection* 19, 221–227.

Cho, C.T. and Moon, B.J. (1984) Studies on the wilt of strawberry caused by *Fusarium oxysporum* f.sp. *fragaria* in Korea. *Korean Journal of Plant Protection* 23, 74–81.

Choma, M.A., Garner, J.L., Marini, R.P. and Barden, J.R. (1982) Effects of fruiting on net photosynthesis and dark respiration of 'Hecker' strawberries. *HortScience* 17, 212–213.

Chung, H.D., Kang, K.Y., Yun, S.J. and Kim, B.Y. (1993) Effect of foliar application of calcium chloride on shelf life and quality of strawberry fruits. *Journal of the Korean Horticulture Society* 24, 7–15.

Civello, P.M., Martínez, G.A., Chaves, A.R. and Añón, M.C. (1995). Peroxidase from strawberry fruit (*Fragaria* × *ananassa* Duch.): Partial purification and determination of some properties. *Journal of Agricultural Food Chemistry* 43, 2596–2601.

Clark, G.A., Albregts, F.L., Stanley, E.E., Smajstrla, A.G. and Zazueta, F.S. (1996) Water requirements and crop coefficients of drip irrigated strawberry plants. *Transactions of the American Society of Agricultural Engineers,* paper no. 90–2532.

Clark, J.H. (1938) Inheritance of the so-called everbearing tendency in the strawberry. *Proceedings of the American Society for Horticultural Science* 35, 67–70.

Collins, J.K. and Perkins-Veazie, P. (1993) Postharvest changes in strawberry fruit stored under simulated retail display conditions. *Journal of Food Quality* 16, 133–143.

Comstock, R.E., Kelleher, T. and Morrow, E.B. (1958) Genetic variation in an asexual species, the garden strawberry. *Genetics* 43, 634–646.

Conroy, R and Ellis, R.F. (1981) Precooling berries improves quality, shortens fluidized freezing time. *Food Processing* 42, 82–83.

Converse, R.H. (1979) Recommended virus-indexing procedures for new USDA small fruit and grape cultivars. *Plant Disease Reporter* 58, 848–851.

Converse, R.H. (1987) Detection and elimination of virus and viruslike diseases in strawberry, In: Converse, R.H. (ed.) *Virus Diseases of Small Fruits.* USDA/ARS. Washington, DC, pp. 2–10.

Converse, R.H. (1992) Modern approaches to strawberry virus research. *Acta Horticulturae* 308, 19–30.

Converse, R.H. and Volk, E. (1990) Some effects of pallidosis disease on strawberry growth under greenhouse conditions. *Plant Disease* 74, 814–816.

Converse, R.H., Denison, W.C. and Lawrence, F.J. (1981) Reaction of some Pacific coast strawberry cultivars to leaf scorch. *Plant Disease* 65, 254–255.

Cooley, D., Schloemann, S. and Tuttle, A. (eds) (1994) *Integrated Pest Management for Strawberries in the Northeastern United States: a Manual for Growers and Scouts.* University of Massachusetts Cooperative Extension System, No. C211, Amherst.

Coop, L.B. and Croft, B.A. (1995) *Neoseiulus fallacis*: dispersal and biological control of *Tetranychus urticae* following minimal inoculations into a strawberry field. *Experimental and Applied Acarology* 19, 31–43.

Cooper, A. (1973) *Root Temperature and Plant Growth.* Commonwealth Agricultural Bureaux, Farnham Royal, UK.

Cooper, A. (1979) *The ABC of NFT.* Grower Books, London. pp. 181.

Couture, R.J., Makhlouf, R.J., Cheour, F. and Willemot, C. (1990) Production of CO_2 and C_2H_2 after gamma irradition of strawberry fruit. *Journal of Food Quality* 13, 385–393.

Craig, D. L. and Aalders, L.E. (1966) Influence of cultural systems on strawberry fruit yield and berry size. *Proceedings of the American Society for Horticultural Science* 89, 318–321.

Craig, D.L. and Brown, C.L. (1977) Influence of digging date, chilling, cultivars and culture on glasshouse strawberry production in Nova Scotia. *Canadian Journal of Plant Science* 57, 571–576.

Craig, D.L., Aalders, L.E., and Bishop, C.J. (1963) The genetic improvement of strawberry progenies through recurrent reciprocal selection. *Canadian Journal of Genetics and Cytology* 5, 33–37.

Craig, D.L., Cutliffe, J.A. and McRae, K.B. (1983) Effect of plant thinning and row width on yield and fruit size of three strawberry cultivars in Atlantic Canada. *Advances in Strawberry Production* 2, 1–6.

Cram, W.T. (1978) The effect of root weevils (Coleoptera: Curculionidae) on yield of five strawberry cultivars in British Columbia. *Journal of the Entomological Society of British Columbia* 75, 10–13.

Cram, W.T. and Neilson, C.L. (1978) Major insect and mite pests of berry crops in British Columbia. *British Columbia Ministry of Agriculture Bulletin* 78–12.

Crane, J.C. and Haut, I.C. (1939) Relationship of width of thinned row to productiveness and quality in the Blakemore strawberry. *Proceedings of the American Society for Horticultural Science* 36, 417–419.

Crock, J.E., Shanks, C.H., and Barritt, B.H. (1982) Resistance in *Fragaria chiloensis* and *F.* × *ananassa* to the aphids *Chaetosiphon fragaefolii* and *C. thomasi. HortScience* 17, 959–960.

Dale, A. (1994) Wild, wild strawberries. *Agri-food Research in Ontario*, March, 8–13.

Dale, A. and Hergert, G.B. (1990) Systems approaches to mechanized harvesting of strawberry crops. In: Dale, A. and Luby, J.J. (eds) *Strawberries into the 20th Century.* Timber Press, Portland, Oregon, pp. 229–236.

Dale, A. and Sjulin, T.M. (1990) Few cytoplasms contribute to North American strawberry cultivars. *HortScience* 25, 1341–1342.

Dale, A., Gray, V.P. and Miles, N.W. (1987) Effects of cultural systems and harvesting techniques on the production of strawberries for processing. *Canadian Journal of Plant Science* 67, 853–862.

Dale, A., Daubeny, H.A., Luffman, M. and Sullivan, J.A. (1992) Development of *Fragaria* germplasm in Canada. *Acta Horticulturae* 348, 75–80.

Dale, A., Hanson, E.J, Yarborough, D.E., McNichol, R.J., Stang, E.J., Brennan, R., Morris, J.R. and Hergert, G.B. (1994) Mechanical harvesting of berry crops. *Annual Review of Horticulture* 14, 255–382.

Daley, L.S., Jahn, O. and Guttridge, C. (1986) Differences between *Fragaria* clones by fourth derivative room temperature spectroscopy of intact leaves. *Photosynthesis Research* 11, 183–188.

Dana, M.N. (1980) The strawberry plant and its environment. In: Childers, N.F. (ed.) *The Strawberry: Cultivars to Marketing.* Horticultural Publications, Gainesville, Florida, pp. 32–44.

Darnell, R.L. and Hancock, J.F. (1996) Balancing vegetative growth with productivity in strawberry. *Proceeding of the IVth North American Strawberry Conference.* University of Florida, Gainesville, pp. 144–150.

Darnell, R.L. and Martin, G.C. (1987) Absorption translocation and metabolism of Et-IAA in relation to fruit set and growth of day-neutral strawberry. *Journal of the American Society for Horticultural Science* 112, 804–807.

Darnell, R.L. and Martin, G.C. (1988) Role of assimilate translocation and carbohydrate accumulation in fruit set of strawberry. *Journal of the American Society for Horticultural Science* 113, 114–118.

Darrow, G.M. (1936) Interrelation of temperature and photoperiodism in the production of fruit buds and runners in the strawberry. *Proceedings of the American Society for Horticultural Science* 34, 360–363.

Darrow, G.M. (1937) Strawberry improvement. In: *Better Plants and Animals 2. USDA Yearbook of Agriculture* , 496–533.

Darrow, G.M. (1953) The Ambato strawberry of Ecuador. *Fruit Varieties and Horticultural Digest* 9, 53–54.

Darrow, G.M. (1955) Leaf variegation in strawberry – a review. *Plant Disease Reporter* 39, 363–370.

Darrow, G.M. (1957a) Report on plant exploration in Chile, Ecuador and Colombia for strawberries and other fruits. *U.S. Department of Agriculture* (mimeo).

Darrow, G.M. (1957b) Exploration in South America for strawberries and other small fruits. *Fruit Varieties and Horticultural Digest* 12, 5–7.

Darrow, G.M. (1966) *The Strawberry. History, Breeding and Physiology.* Holt, Rinehart and Winston, New York.

Darrow, G.M. and Dewey, G.W. (1934) Studies on stomata of strawberry varieties and species. *Proceedings of the American Society for Horticultural Science* 32, 440–447.

Darrow, G.M. and Scott, D.H. (1947) Breeding for cold hardiness of strawberry flowers. *Proceedings of the American Society for Horticultural Sciences* 50, 239–242.

Darrow, G.M. and Sherwood, H. (1931) Transpiration studies in strawberries. *Proceedings of the American Society for Horticultural Sciences* 28, 225–230.

Darrow, G.M. and Waldo, G.F. (1934) Responses of strawberry varieties and species to the duration of the daily light period. *USDA Technical Bulletin 453*, Washington DC.

Darrow, G.M., Waldo, G.F. and Schuster, C.E. (1933) Twelve years of strawberry breeding: a summary of the strawberry breeding work of the United States Department of Agriculture. *Journal of Heredity* 24, 391–402.

Darrow, G.M., Wilcox, M.S., Scott, D.H. and Hutchins, M.C. (1947) Breeding strawberries for vitamin C. *Journal of Heredity* 38, 363–365.

Darrow, G.M., Scott, D.H., and Goheen, A.C. (1954) Relative resistance of strawberry varieties to powdery mildew at Beltsville, Maryland. *Plant Disease Reporter* 38, 864–866.

Daubeny, H.A. (1961) Powdery mildew resistance in strawberry progenies. *Canadian Journal of Plant Science* 41, 239–243.

Daubeny, H.A. (1963) Effect of parentage in breeding for red stele resistance of strawberry in British Columbia. *Proceedings of the American Society for Horticultural Science* 84, 289–294.

Daubeny, H. A. (1990) Strawberry breeding in Canada. *HortScience* 25, 893–994.

Daubeny, H.A. and Pepin, H.S. (1965) The relative resistance of various *Fragaria chiloensis* clones to *Phytophthora fragariae*. *Canadian Journal of Plant Science* 45, 365–368.

Daubeny, H.A. and Pepin, H.S. (1977) Evaluation of strawberry clones for fruit rot resistance. *Journal of the American Society for Horticultural Science* 102, 431–435.

Daubeny, H.A. and Anderson, A. (1993) Prospects for new fresh market strawberry cultivars for the Pacific Northwest. *Acta Horticulturae* 348, 177–179.

Daubeny, H.A., Norton, R.A., Schwartze, C.D. and Barritt, B.H. (1970) Winter-hardiness in strawberries for the Pacific Northwest. *HortScience* 5, 152–154.

Davis, M.B., Hill, H. and Johnson, F.B. (1934) Nutritional studies with *Fragaria*. II. A study of the effect of deficient and excess K, P, Mg, Ca and S. *Scientific Agriculture* 14, 411–432.

Davis, T. (1995) Use of molecular markers to accelerate cultivar development in *Fragaria*. *Proceedings of the IV North American Strawberry Conference*, Orlando, Florida, pp. 113–118.

Davis, T.M. and Yu, H. (1997) A linkage map of the diploid strawberry, *Fragaria vesca*. *Journal of Heredity* 88, 215–221.

Day, D. (1993) *Strawberries*. Grower Digest Series No. 3. Nexus Business Communications, London.

Day, W.H., Hedlund, R.C., Saunders, L.B. and Coutinot, D. (1990) Establishment of *Peristenus digoneutis* (Hymenoptera: Braconidae), a parasite of the tarnished plant bug (Hemiptera: Miridae), in the United States. *Environmental Entomology* 19, 1528–1533.

Decou, G.C. (1994) Biological control of the two-spotted spider mite (Acarina: Tetranychidae) on commercial strawberries in Florida with *Phytoseiulus persimilis* (Acarina: Phytoseiidae). *Florida Entomological Society* 77, 33–41.

Degani, C., Rowland, L.J., Levi, A., Hortynski, J.A. and Galletta, G.J. (1998) DNA fingerprinting of strawberry (*Fragaria × ananassa*) cultivars using randomly amplified polymorphic DNA (RAPD) markers. *Euphytica* 102, 247–253.

Delhomez, N., Carisse, O., Lareau, M., and Khanizadeh, S. (1995) Susceptibility of strawberry cultivars and advanced selections to leaf spot caused by *Mycosphaerella fragariae*. *HortScience* 30, 592–595.

Delp, B.R. and Milholland, R.S. (1981) Susceptibility of strawberry cultivars and related species to *Colletotrichum fragariae*. *Plant Disease* 65, 421–423.

De Moesbach, E.W. (1992) *Botánica Indigena de Chile*. Edit. Andrés Bello, Santiago, Chile.

Dénes, F. (1997) Performance of new strawberry cultivars under Hungarian conditions. *Acta Horticulturae* 439, 285–290.

Deng, H. and Ueda, Y. (1993) Effects of freezing methods and storage temperature on flavor stability and ester contents of frozen strawberries. *Journal of the Japanese Horticulture Society* 62, 633–639.

Denison, E.L. and Buchele, W.F. (1967) Mechanical harvesting of strawberries. *Proceedings of the American Society for Horticultural Science* 91, 267–273.

Dennis, C. and Mountford, J. (1975) The fungal flora of soft fruits in relation to storage and spoilage. *Annals of Applied Biology* 79, 141–147.

Denoyes, B. and Baudry, A. (1995) Species identification and pathogenicity study of French *Colletotrichum* strains isolated from strawberry using morphological and cultural characteristics. *Phytopathology* 85, 53–57.

Denoyes-Rothan, B. (1997) Inheritance of resistance to *Colletotrichum acutatum* in strawberry (*Fragaria × ananassa*). *Acta Horticulturae* 439, 809–814.

Denoyes-Rothan, B. and Guérin, G. (1996) Comparison of six inoculation techniques with *Colletotrichum acutatum* on cold stored strawberry plants and screening for resistance to this fungus in French strawberry collections. *European Journal of Plant Pathology* 102, 615–621.

D'Ercole, N., Niopti, P., Finessi, L.E. and Manzali, D. (1988) Results of several years' research in Italy of biological control of soil fungi with *Trichoderma* ssp. *Bulletin of the OEPP* 18, 95–102.

Dicklow, M.B., Acosta, N. and Zuckerman, B.M. (1993) A novel *Streptomyces* for controlling plant-parasitic nematodes. *Journal of Chemical Ecology* 19, 159–173.

Dirinck, P.J., De Pooter, H.L., Willaert, G.A. and Schamp, N.M. (1981) Flavor of cultivated strawberries: the role of the sulfur compounds. *Journal of the Science of Food Agriculture* 29, 316–321.

Dogadkina, N.A. (1941) A contribution to the question of genome relations in some species of *Fragaria*. *Compt. Rend. (Doklady) Acad. Sci. USSR* 30, 166–168.

Dorn, R. D. (1984) *Vascular Plants of Montana*. Mountain West Publishers, Cheyenne, Wyoming.

Doss, R.P. and Shanks, C.H. (1988) The influence of leaf pubescence on the resistance of clones of beach strawberry (*Fragaria chiloensis* [L.] Duchesne) to adult black vine weevils (*Otiorhynchus sulcatus* F.). *Scientia Horticulturae* 34, 47–54.

Doss, R.P., Shanks, C.H., Chamberlain, J.D. and Garth, J.K.L. (1987) Role of leaf hairs in resistance of a clone of beach strawberry, *Fragaria chiloensis*, to feeding by adult black vine weevil, *Otiorhynchus sulcatus* F. (Coleoptera: curculionidae). *Environmental Entomology* 16, 764–768.

Douillard, C. and Guichard, E. (1990) The aroma of strawberry (*Fragaria ananassa*): characterization of some cultivars and influence of freezing. *Journal of the Science of Food and Agriculture* 50, 517–531.

Downs, R.J. and Piringer, A.A. (1955) Differences in photoperiod reponses of everbearing and June-bearing strawberries. *Proceedings of the American Society for Horticultural Science* 66, 234–236.

Draper, A.D., Galletta, G.J. and Swartz, H.J. (1981) 'Tribute' and 'Tristar' everbearing strawberries. *HortScience* 16, 794–795.

Dreher, T.W. and Poovaiah, B.W. (1982) Changes in auxin content during development in strawberry fruits. *Journal of Plant Growth and Regulation* 1, 267–276.

Duchesne, A. N. (1766) *Histoire Naturelle des Fraisiers*, Paris.

Duewer, R.G. and Zych, C.C. (1967) Heritability of soluble solids and acids in progenies of the cultivated strawberry (*Fragaria* × *ananassa* Duch.). *Proceedings of the American Society for Horticultural Science* 90, 153–157.

Duncan, J.M. (1980) A technique for detecting red stele (*Phytophthora fragariae*) infection of strawberry stocks before planting. *Plant Disease* 64, 1023–1025.

Duncan, J.M. and Kennedy, D.M. (1995) Effect of temperature and host genotype on the production of inoculum by *Phytophthora fragariae* var. *fragariae* from the roots of infected strawberry plants. *Plant Pathology* 44, 10–21.

Dunne, M.J., and Fitter, A.H. (1989) The phosphorus budget of a field-grown strawberry (*Fragaria* × *ananassa* cv Hapil): evidence for a mycorrhizal contribution. *Annals of Applied Biology* 114, 185–193.

Du Pressis, H.J., Brand, R.J., Glynn-Woods, C. and Goedhart, M.A. (1995) Genetic engineering leads to a herbicide-tolerant strawberry. *South African Journal of Science* 91, 218.

Durner, E.F. and Poling, E.B. (1988) Strawberry developmental responses to photoperiod and temperature. *Advances in Strawberry Production* 7, 6–15.

Durner, E.F., Barden, J.A., Himelrick, D.G. and Poling, E.B. (1984) Photoperiod and temperature effects on flower and runner development in day-neutral, June-bearing and everbearing strawberries. *Journal of the American Society for Horticultural Science* 109, 396–400.

Durner, E. F., Barden, J.A. and Himelrick, D.H. (1985) Net photosynthesis, growth and development of three photoperiod types of strawberries as effected by photoperiod. *Advances in Strawberry Production* 4, 28–31.

Durner, E.F., Poling, E.B. and Albregts, E.A. (1986) Early season yield responses of selected strawberry cultivars to photoperiod and chilling in a Florida winter production system. *Journal of the American Society for Horticultural Science* 112, 53–56.

Dwyer, L.M., Stewart, D.W., Houwing, L. and Balchin, D. (1987) Response of strawberries to irrigation scheduling. *HortScience* 22, 42–44.

Eastburn, D.M. and Gubler, W.D. (1992) Effects of soil moisture and temperature on the survival of *Colletotrichum acutatum. Plant Disease* 76, 841–842.

Easterbrook, M.A., Crook, A.M.E., Cross, J.V. and Simpson, D.W. (1997) Progress towards integrated pest management on strawberry in the United Kingdom. *Acta Horticulturae* 439, 899–904.

Eaves, C.A. and Leefe, J.S. (1962) Note on the influence of foliar sprays of Ca on the firmness of strawberries. *Canadian Journal of Plant Science* 42, 746–747.

Edwards, W.H., Jones, R.K. and Schmitt, D.P. (1985) Host suitability and parasitism of selected strawberry cultivars by *Meloidogyne hapla* and *M. incognita. Plant Disease* 69, 40–42.

El-Farhan, A.H. and Pritts, M.P. (1997) Water requirements and water stress in strawberry. *Advances in Strawberry Research* 16, 5–12.

El Ghaouth, C., Arul, J., Ponnampalam, R. and Boulet, M. (1991) Chitosan coating effect on the storability and quality of fresh strawberries. *Journal of Food Science* 56, 1618–1620.

El-Kazzaz, M.K., Sommer, N.F. and Fortlage, R.J. (1983) Effect of different atmospheres on postharvest decay and quality of fresh strawberries. *Phytopathology* 73, 282–285.

Ellis, J.R. (1958) Intergeneric hybridization between *Fragaria* and *Potentilla. International Genetics Congress*, Vol. II, p. 74.

Ellis, J. R. (1962) *Fragaria – Potentilla* intergeneric hybridization and evaluation in *Fragaria. Proceedings of the Linnean Society of London* 173, 99–106.

Ellis, J.R. (1963) The allopolyploid hybrid between *Fragaria vesca* and *F. moschata. Proceedings of the 11th International Congress of Genetics* 1, 209–210.

Ellis, M.A. (1995) New directions in research on fruit fungal pathogens. In: Pritts, M.P., Chandler, C.K. and Crocker, T.E. (eds) *Proceedings of the IV North American Strawberry Conference*, University of Florida, Gainesville, pp. 42–47.

Ellis, M.A., Wilcox, W.F. and Madden, L.V. (1998) Efficacy of metalaxyl, fosetyl-aluminum and straw mulch for the control of strawberry leather rot, caused by *Phytophthora cactorum. Plant Disease* 82, 329–333.

Eshenaur, B.C. and Milholland, R.D. (1989) Factors influencing the growth of *Phomopsis obscurans* and disease development on strawberry leaf and runner tissue. *Plant Disease* 73, 814–819.

Evans, W. D. (1974) Evidence of a crossibility barrier in diploid × octoploid crosses in the genus *Fragaria*. *Euphytica* 23, 95–100.

Evans, W. D. (1977) The use of synthetic octoploids in strawberry breeding. *Euphytica* 26, 497–503.

Evans, W. D. (1982a) Guelph SO1 synthetic octoploid strawberry breeding clone. *HortScience* 17, 833–834.

Evans, W. D. (1982b) Guelph SO2 synthetic octoploid strawberry breeding clone. *HortScience* 17,834.

Evans, W.D. and Jones, J.K. (1967) Incompatibility in *Fragaria*. *Canadian Journal of Genetics and Cytology* 9, 831–836.

Faby, R. (1997) The productivity of graded 'Elsanta' frigo plants from different regions. *Acta Horticulturae* 439, 449–455.

Fadeeva, T. S. (1966) Communication 1. Principles of genome analysis (with reference to the genus *Fragaria*). *Genetika* 2, 6–16.

Faedi, W., Baruzzi, G., Dradi, R., Rosati, P. and Lucci, P. (1997) Strawberry breeding in Italy. *Acta Horticulturae* 439, 121–128.

Fallik, E., Archbold, D.D., Hamilton-Kemp, T.R., Clements, A.M., Collins, R.W. and Barth, M.M. (1998) (E)-2-Henenal can stimulate *Botrytis cinerea* growth *in vitro* and on strawberry *in vivo* during storage. *Journal of the American Society for Horticultural Science* 123, 875–881.

Fear, C.D. and Nonnecke, G.R. (1989) Soil mulches influence reproductive and vegetative growth of 'Fern' and 'Tristar' day-neutral strawberries. *HortScience* 6, 912–913.

Federova, N.J. (1946) Crossibility and phylogenetic relations in the main European species of *Fragaria*. *Compt. Rend. (Doklady) Acad. Sci. USSR*. 30, 545–547.

Ferree, D.C. and Stang, E.J. (1988) Seasonal plant shading, growth and fruiting in 'Earliglow' strawberry. *Journal of the American Society for Horticultural Sciences* 113, 322–327.

Ferreira, M.D., Bartz, J.A., Sargent, S.A. and Brecht, J.K. (1996) An assessment of the decay hazard associated with hydrocooling strawberries. *Plant Disease* 80, 1117–1122.

Ferreira, M.D., Brecht, J.K., Sargent, S.A. and Aracena, J.J. (1994) Physiological responses of strawberry to film wrapping and precooling methods. *Proceedings of the Florida State Horticulture Society* 107, 265–268.

Finn, C., Hancock, J. and Heider, C. (1998) Notes on the strawberry of Ecuador: ancient land races, the community of farmers and modern production. *HortScience* 33, 583–587.

Fiola, J. (1996) Advances in strawberry plant production. *Proceedings of the IV North American Strawberry Conference*. University of Florida, Gainesville, pp. 180–191.

Fishman, R. (1987) Albert Etter: fruit breeder. *Fruit Varieties Journal* 41, 40–46.

Fletcher, S.W. (1917) *The Strawberry in North America: History, Origin, Botany and Breeding*. The Macmillan Company, New York.

Flore, J.A. and Lakso, A. (1989) Environmental and physiological regulation of photosynthesis in fruit crops. *Horticulture Reviews* 11, 111–157.

Forney, C.F. and Breen, P.J. (1985) Dry matter partitioning and assimilation in fruiting and deblossomed strawberry. *Journal of the American Society for Horticultural Science*. 110, 181–185.

Forney, C.F. and Breen, P.J. (1986) Sugar content and uptake in the strawberry fruit. *Journal of the American Society for Horticultural Science* 111, 241–247.

Frazier, N.W. (1974) Six new strawberry indicator clones evaluated for the detection and diagnosis of twelve graft-transmissible diseases. *Plant Disease Reporter* 58, 28–31.

Frazier, N.W., Voth, V. and Bringhurst, R.S. (1965) Inactivation of two strawberry viruses in plants grown in a natural high temperature environment. *Phytopathology* 55,1203–1205.

Funt, R.C., Ellis, M.A. and Wally, C. (1997) *Midwest Small Fruit Pest Management Handbook.* Bulletin 861, Ohio University, Columbus.

Gagnon, B., Desjardins, Y. and Bedard, R. (1990) Fruiting as a factor in accumulation of carbohydrates and nitrogen and in fall hardening of day neutral strawberries. *Journal of the American Society for Horticultural Science* 115, 520–525.

Galletta, G. J. and Bringhurst, R.S. (1990) Strawberry management. In: Galletta, G.J. and Himelrick, D. (eds) *Small Fruit Crop Management.* Prentice Hall, Englewood Cliffs, New Jersey, pp. 83–156.

Galletta, G.J. and Maas, J.L. (1990) Strawberry genetics. *HortScience* 25,871–879.

Galletta, G.J., Draper, A.D. and Scott, D.H. (1981a) The US Department of Agriculture strawberry breeding program. *HortScience* 16, 743–746.

Galletta, G.J., Draper, A.D. and Swartz, H.J. (1981b) New everbearing strawberries. *HortScience* 16, 726.

Galletta, G.J., Draper, A.D. and Maas, J.L. (1989) Combining disease resistance, plant adaptation and fruit quality in breeding short day and day-neutral strawberries. *Acta Horticulturae* 265, 43–51.

Galletta, G.J., Smith, B.J. and Gupton, C.L. (1993) Strawberry parent clones US 70, US 159, US 292, and US 438 resistant to *Anthracnose* crown rot. *HortScience* 28, 1055–1056.

Galletta, G.J., Maas, J.L., Finn, C.E., Smith, B.J., Gupton, C.L., Luby, J.J. and Wildung, D.K. (1997) New strawberries from the USDA cooperative breeding programs. *Acta Horticulturae* 439, 227–232.

Gamliel, A. and Stapleton, J.J. (1993) Characterization of antifungal volatile compounds evolved from solarized soil amended with cabbage residues. *Phytopathology* 83, 899–905.

Garcia, J.M., Aguilera, C. and Miguel, A. (1995) Post harvest heat treatment on spanish strawberry (*Fragaria* × *ananassa* cv. Tudla). *Journal of Agricultural Food Chemistry* 43, 1489–1492.

Garcia, J.M., Herrera, S. and Morilla, A. (1996a) Effects of post-harvest dips in calcium chloride on strawberry. *Journal of Agricultural Food Chemistry* 44, 30–33.

Garcia, J.M., Herrera, S. and Morilla, A. (1996b) Grey mold in and quality of strawberry fruit following post-harvest heat treatment. *HortScience* 31, 255–257.

Gast, K. L. B. and Pollard, J.E. (1991) Rowcovers enhance reproductive and vegetative yield components in strawberries. *HortScience* 26, 1467–1469.

Gehrmann, A. (1995) Growth, yield and fruit quality of strawberries as affected by water supply. *Acta Horticulturae* 171, 463–469.

Gent, M.P.N. (1990) Ripening and fruit weight of eight strawberry cultivars respond to row cover removal date. *Journal of the American Society for Horticultural Science* 115, 202-207.

Giovanardi, R. and Testolin, R. (1984) Evapotranspiration and yield response of strawberry (*Fragaria* × *ananassa* Duch.) as affected by soil water conditions. *Irrigazione* 4, 15–25.

Given, N.K., Venis, M.A. and Grierson, D. (1988a) Hormonal regulation of ripening in the strawberry, a non-climacteric fruit. *Planta* 174, 402–406.

Given, N.K., Venis, M.A. and Grierson, D. (1988b) Purification and properties of phenylalanine ammonia-lyase from strawberry fruit and its synthesis during ripening. *Journal of Plant Physiology* 133, 31–37.

Given, N.K., Venis, M.A. and Grierson, D. (1988c) Phenylalanine ammonia-lyase activity and anthocyanin synthesis in ripening strawberry fruit. *Journal of Plant Physiology* 133, 25-30.

Gleason, M.L., Nonnecke, G.R. and Fear, C.D. (1989) Plant growth and yield of day-neutral and June-bearing strawberry cultivars in response to soil fumigation. *Advances in Strawberry Production* 8, 51–52.

Glenn, D.M. and Takeda, F. (1989) Guttation as a technique to evaluate the water status of strawberries. *HortScience* 24, 599–601.

Goff, L.M. (1986) Certification of horticultural crops – a state perspective. In: Zimmerman, R.H., Griesbach, R.J., Hammerschlag, F.A. and Lawson, R.H. (eds) *Tissue Culture as a Propagation System for Horticultural Crops*. Martinus Nijhoff, The Netherlands, pp. 139–146.

Gonzalez de Nájera, A. (1866) Desengaño y reparo de la guerra del Reino de Chile. *Colección de documentos inéditos para la historia de España. Tomo XLVIII.* Madrid, Imprenta de la Viuda de Calero.

González-Zamora, Ribes, A., Mesequer, A. and García-Marí, F. (1994) Thrips control in strawberry: use of broad beans as a refuge for populations of anthocorids. *Boletín Sanidad Vegetal Plagas* 20, 57–72.

Gooding, H.J. (1972) Studies on field resistance of strawberry varieties to *Phytophthora fragariae*. *Euphytica* 21, 63–70.

Gooding, H.J., Jennings, D.L. and Topham, T.P. (1975) A genotype-environmental experiment on strawberries in Scotland. *Heredity* 34,105–115.

Gooding, H.J., McNicol, R.J. and MacIntyre, D. (1981) Methods of screening strawberries for resistance to *Sphaerotheca macularis* (Wall ex Frier) and *Phytophthora cactorum* (Leb. and Cohn). *Journal of Horticultural Science* 56, 239–245.

Gooding, H.J., McNicol, R.J. and Reid, J.H. (1983) Studies of strawberry fruit characteristics necessary for machine harvesting. *Crop Research* 23, 3–16.

Goodman, R.D. and Oldroyd, B.P. (1988) Honeybee pollination of strawberries (*Fragaria* × *ananassa* Duchesne). *Australian Journal of Experimental Agriculture* 28, 435–438.

Goulart, B.L. and Demchak, K. (1994) Cryoprotectants prove ineffective for frost protection on strawberries. *Journal of Small Fruit and Viticulture* 2, 45–52.

Govorova, G. (1992) Strawberry breeding in Russia. *Acta Horticulturae* 348, 45–55.

Graham, J, McNicol, R.J. and Greig, K. (1995) Towards genetic based insect resistance in strawberry using the Cowpea trypsin inhibitor gene. *Annals of Applied Biology* 127, 163-173.

Graham, J., McNichol, R.J. and McNicol, J.W. (1996) A comparison of methods for the estimation of genetic diversity in strawberry cultivars. *Theoretical and Applied Genetics* 93, 402–406.

Graham, J., Gordon, S.C. and McNicol, R.J. (1997) The effect of the CpTi gene in strawberry against attack by vine weevil (*Otiorhynchus sulcatus* F. Coleoptera: Curculionidae). *Annals of Applied Biology* 131, 133–139.

Graichen, K., Kegler, H. and Ulrich, M. (1985) Zur Toleranz der Erdbeere gegenuber der Krauselkrankheit. *Archiev für Gartenbau (Berlin)* 33, 275–283.

Green, A. (1971) Soft fruits. In: Hulme, A.C. (ed.) *The Biochemistry of Fruits and their Products*, Vol. 2. Academic Press, London, pp. 375–409.

Gruda, Z. and Kurzeba, W. (1981) A method of and apparatus for the continuous freezing of food products in bulk. *US Patent* no. 4283923.

Grunfeld, E., Vincent, C. and Bagnanas, D. (1989) High-performance liquid chromatography analysis of nectar and pollen of strawberry flowers. *Journal of Agricultural Food Chemistry* 37, 290–294.

Gubler, W.D., Nelson, M.D. and Feliciano, A. (1995) New directions in research on foliar fungal pathogens. IV. North American Strawberry Conference. Orlando, Florida, pp. 39–41.

Gunnell, P.S. and Gubler, W.D. (1992) Taxonomy and morphology of *Colletotrichum* species pathogenic to strawberry. *Mycologia* 84, 157–165.

Gupton, C.L. and Smith, B.J. (1991) Inheritance of resistance to *Colletotrichum* species in strawberry. *Journal of the American Society for Horticultural Science* 116, 724–727.

Guttridge, C.G. (1955) Observations on the shoot growth of the cultivated strawberry plant. *Journal of Horticultural Science* 30, 1–11.

Guttridge, C.G. (1985) *Fragaria* × *ananassa*. In: Halvey, A. H. (ed.) *CRC Handbook of Flowering*, Vol. III. CRC Press, Boca Raton, Florida, pp. 16–33.

Guttridge, C.G. and Mason, D.T. (1966) Effects of post-harvest defoliation of strawberry plants on truss initiation, crown branching and yield. *Horticultural Research* 6, 22–32.

Guttridge, C.G. and Turnbull, J.M. (1974) Improving anther dehiscence and pollen germination in strawberry with boric acid and salts of divalent cations. *Horticultural Research* 14, 73- 79

Guttridge, C.G., Mason, D.T. and Ing, E.G. (1965) Cold storage of strawberry runner plants at different temperatures. *Experimental Horticulture* 12, 38–41.

Hamilton-Kemp, T.R., Anderson, R.A., Rodrequez, J.G., Loughrin, J.H. and Patterson, C.G. (1988) Strawberry foliage headspace vapor components at periods of susceptibilty and resistance to *Tetranychus urticae. Journal of Chemical Ecology* 14, 789–796.

Hancock, J. F. (1987) Are ribbons superior to matted rows? A review of research supported by the North American Strawberry Growers Association. *Advances in Strawberry Production* 6, 53–54.

Hancock, J.F. (1990) Ecological genetics of natural strawberry species. *HortScience* 25, 869–870.

Hancock, J.F. (1992) *Plant Evolution and the Origin of Plant Species*. Prentice-Hall, Englewood Cliffs, New Jersey.

Hancock, J. F. and Bringhurst, R.S. (1979a) Hermaproditism in predominately dioecious populations of *Fragaria chiloensis* (L.) Duch. *Bulletin of the Torrey Botanical Club* 106, 229–231.

Hancock, J.F. and Bringhurst, R.S. (1979b) Ecological differentiation in perennial, octoploid species of *Fragaria. American Journal of Botany* 66, 367–375.

Hancock, J.F. and Bringhurst, R.S. (1980) Sexual dimorphism in the strawberry *Fragaria chiloensis. Evolution* 34, 762–768.

Hancock, J.F. and Bringhurst, R.S. (1988) Yield component interactions in wild populations of California *Fragaria. HortScience* 23, 889–890.

Hancock, J. F. and Cameron, J.S. (1986) Effect of harvesting in the first year on yield and dry matter partitioning in strawberry. *Advances in Strawberry Production* 5, 7–11.

Hancock, J. F. and Luby, J.J. (1993) Genetic resources at our doorstep : the wild strawberries. BioScience 43, 141–147.

Hancock, J.F., and Luby, J.J. (1995) Adaptive zones and ancestry of the most important North American strawberry cultivars. *Fruit Varieties Journal* 49, 85–89.

Hancock, J.F. and Simpson, D. (1995) Methods of extending the strawberry season in Europe. *HortTechnology* 5,286–290.

Hancock, J. F., Siefker, J.H. and Schulte, N.L. (1983) Cultivar variation in yield components of strawberries. *HortScience* 18,312–313.

Hancock, J. F., Pritts, M.P. and Siefker, J.H. (1984a) Yield components of strawberries maintained in ribbons and matted rows. *Crop Research* 24, 37–43.

Hancock, J. F., Moon, J.W. and Flore, J.A. (1984b) Within-row spacing and dry weight distribution in two strawberry cultivars. *HortScience* 19,412–413.

Hancock, J.F., Flore, J.A and Galletta, G.J. (1989a) Gas exchange properties of strawberry species and their hybrids. *Scientia Horticulturae* 40, 139–144.

Hancock, J.F., Flore, J.A. and Galletta, G.J. (1989b) Variation in leaf photosynthetic rates and yield in strawberries. *Journal of the Society for Horticultural Science* 64, 449–454.

Hancock, J. F., Maas, J.L., Shanks, C.H., Breen, P.J. and Luby, J.J. (1990) Strawberries (*Fragaria* ssp). In: Moore, J. N. and Ballington, J. (eds) *Genetic Resources in Temperate Fruit and Nut Crops.* International Society of Horticultural Sciences, Wageningen, The Netherlands, pp. 489–546.

Hancock, J. F., Dale, A. and Luby, J. (1993) Should we reconstitute the strawberry. *Acta Horticulturae* 348, 85–93.

Hancock, J. F., Callow, P. and Shaw, D. (1994) Randomly amplified polymorphic DNA (RAPDs) in the cultivated strawberry *Fragaria × ananassa. Journal of the American Society for Horticultural Science* 119, 862–864.

Hancock, J.F., Luby, J.J., Dale, A. and Darnell, D.L. (1996a) Germplasm resources in octoploid strawberries: Potential sources of genes to increase yields in northern climates. *Proceedings of the IV North American Strawberry Conference*, University of Florida, Gainesville, pp. 87–94.

Hancock, J. F., Scott, D.H. and Lawrence, F.J. (1996b) Strawberries. In: Janick, J. and Moore, J.N. (eds) *Fruit Breeding,*Vol. II, *Vine and Small Fruits.* John Wiley & Sons, New York, pp. 419–470.

Hancock, J.F., Finn, C. and Heider, C. (1997) A farmer based approach to conserve the historic Andean strawberry. *Chronica Horticulturae* 37, 14–16.

Hancock, J.F., Luby, J.J., Goulart, B.L. and Pritts, M.P. (1998) The strawberry matted row: practical cropping system or dated anachronism? *Advances in Strawberry Research* 16, 1–4.

Hancock, J.F., Lavín, A. and Retamales, J.B. (1999) Our southern strawberry heritage: *Fragaria chiloensis* of Chile. *HortScience* (in press).

Handley, D.T. and Pollard, J.E. (1993) Microscopic examination of tarnished plant bug

(Heteroptera: Miridea) feeding damage to strawberry. *Journal of Economic Entomology* 86, 505–510.

Handley, D.T., Dill, J.F. and Pollard, J.E. (1991) Field susceptibility of twenty strawberry cultivars to tarnished plant bug injury. *Fruit Varieties Journal* 45, 166–169.

Hansche, P.E., Bringhurst, R.S. and Voth, V. (1968) Estimates of genetic and environmental parameters in the strawberry. *Proceedings of the American Society for Horticultural Science* 92, 338–345.

Hansen, C.M. and Ledebuhr, R.L. (1980) Mechanical harvesting of strawberries in Michigan. In: *Agricultural Experiment Station Bulletin 645*, Oregon State University, Portland, pp. 54-62.

Hansen, E. and Waldo, C.F. (1944) Ascorbic acid content of small fruits in relation to genetic and environmental factors. *Food Research* 9, 453–461.

Hanson, H.C. (1931) Comparison of root and top development in varieties of strawberry. *American Journal of Botany* 18, 658–673.

Harborne, J.C. (1984) Anthocyanins. In: *Phytochemical Methods*. Chapman & Hall, London, p. 64.

Hardenburg, R.E., Watada, A.E. and Wang, C.Y. (1986) *The Commercial Storage of Fruits, Vegetables, and Florist and Nursery Stocks*. USDA Agricultural Handbook 66, Washington, DC.

Harland, S.C. and King, E. (1957) Inheritance of mildew resistance in *Fragaria* with special reference to cytoplasmic effects. *Heredity* 11, 257.

Harris, C.M. and Harvey, J.M. (1973) Quality and decay of California strawberries stored in CO_2-enriched atmospheres. *Plant Disease Reporter* 57, 44–46.

Harris, R. E. (1973) Relative hardiness of strawberry cultivars at three times of the winter. *Canadian Journal of Plant Science* 53, 147–152.

Harris, J.E. and Dennis, C. (1980) Distribution of *Mucor piriformis*, *Rhizopus sexualis* and *R. stolonifer* in relation to spoilage of strawberries. *Transactions of the British Mycological Society* 75, 445–450.

Harris, D.C. and Yang, J.R. (1996) The relationship between the amount of *Verticillium dahliae* in soil and the incidence of strawberry wilt as a basis for disease risk prediction. *Plant Pathology* 45, 106–114.

Harrison, R.E., Luby, J.J., Furnier, G.R., Hancock, J.F. and Cooley, D. (1996) Variation for crown rot and powdery mildew susceptibility in wild strawberry from North America. *Eucarpia Fruit Breeding and Genetics Symposium*, Oxford.

Harrison, R.E., Luby, J.J. and Furnier, G.R. (1997a) Chloroplast DNA restriction fragment variation among strawberry taxa (*Fragaria* spp.). *Journal of the American Society for Horticultural Science* 122, 63–68.

Harrison, R.E., Luby, J.J., Furnier, G.R. and Hancock, J.F. (1997b) Morphological and molecular variation among populations of octoploid *Fragaria virginiana* and *F. chiloensis* (Rosaceae) from North America. *American Journal of Botany* 84, 612–620.

Hartmann, H.T. (1947a) Some effects of temperature and photoperiod on flower formation and runner production in the strawberry. *Plant Physiology* 22, 401–420.

Hartmann, H.T. (1947b) The influence of temperature on the photoperiodic response of strawberries varieties grown under controlled environmental conditions. *Proceedings of the American Society for Horticultural Science* 50, 243–245.

Hartz, T.K., DeVay, J.E. and Elmore, C.L. (1993) Solarization is an effective soil disinfestation technique for strawberry production. *HortScience* 28, 104–106.

Harvey, J.M. (1982) CO_2 atmospheres for truck shipments of strawberries. In: Richardson, D.G. and Meheriuk, M. (eds) *Controlled Atmospheres for Storage and Transport of Perishable Agricultural Commodities.* Timber Press, Portland, Oregon, pp. 359–365.

Harvey, J.M., Harris, C.M., and Porter, F.M. (1971) *Air Transport of California Strawberries: Pallet Covers to Maintain Modified Atmospheres and Reduce Market Losses. USDA* Marketing Research Report no. 920, Washington, DC.

Havis, L. (1938) Freezing injury to strawberry flower buds, flowers and young fruits. *Ohio Agriculture Experiment Station Bimonthly Bulletin* 23, 168–172.

Haymes, K.M., Van de Weg, W.E., Arens, P., Vosman, B. and Den Nijs, A.P.M. (1997) Molecular mapping and construction of scar markers of the strawberry Rpf1 resistance gene to *Phytophthora fragariae* in strawberry. *Acta Horticulturae* 439, 845–851.

Haymes, K.M., Henken, B., Davis, T.M. and van de Weg, W.E. (1997) Identification of RAPD markers linked to a *Phytophthora fragariae* resistance gene (Rpf1) in the cultivated strawberry. *Theoretical and Applied Genetics* 93, 1202–1210.

Hedlund, R.C. and Graham, H.M. (1987) *Economic Importance and Biological Control of Lygus and Adelphocoris in North America.* ARS–64. Agricultural Research Service, United States Department of Agriculture, Washington, DC.

Hedrick, U.P. (1925) *The Small Fruits of New York.* J.B. Lyon, Albany, New York.

Heide, O.M. (1977) Photoperiod and temperature interactions in growth and flowering of strawberry. *Physiologia Plantarum* 40, 21–26.

Hellman, E.W. and Travis, J.D. (1988) Growth inhibition of strawberry at high temperature. *Advances in Strawberry Production* 7, 36–39

Hemphill, D.D. (1990) Pest management systems for strawberry weeds. In: Pimentel, E. (ed.) *Handbook of Pest Management in Agriculture.* CRC Press, Boca Raton, Florida, pp. 405–415.

Hemphill, R. and Martin, L.H. (1992) Microwave oven-drying method for determining soluble solids in strawberries. *HortScience* 27,1326.

Hergert, G.B. and Dale, A. (1989) Performance of a mechanized strawberry production system. ASAE/CSAE meeting presentation, no. 89–1974.

Hiertaranta, T. and Linna, M. (1997) Selecting cultivars for strawberry production in Finland–clone and cultivar trials. *Acta Horticulturae* 439, 145–148.

Hildreth, A.C. and Powers, L. (1941) The Rocky Mountain strawberry as a source of hardiness. *Proceedings of the American Society for Horticultural Science* 38, 410–412.

Hill, R.G. and Haut, I.C. (1949) Growth and yield responses of the Temple strawberry as influenced by plant spacing, width of row and renewal system. *Proceedings of the American Society for Horticultural Science* 54, 192–196.

Himelrick, D.G. and Dozier, W.A. (1991) Soil fumigation and soil solarization in strawberry production. *Advances in Strawberry Production* 10, 12–28.

Hirvi, T. (1983) Mass fragmentographic and sensory analysis in the evaluation of the aroma of some strawberry varieties. *Lebensmittel-Wissenschaft Technologie* 16, 157–161.

Hirvi, T. and Honkanen, E. (1982) The volatiles of two new strawberry cultivars, Annelie and Alaska Pioneer obtained by backcrossing of cultivated strawberries with wild strawberries, *Fragaria vesca*, Rügen and *Fragaria virginiana. Zeitschrift fur Lebensmittel-Untersuchung und Forschung* 175, 113–116.

Hochmuth, G.J. and Albregts, E.E. (1994) Fertilization of strawberries in Florida. *University of Florida Cooperative Extension Service*, Circular 1141, 1–4.

Hochmuth, G.J., Albregts, E.E., Chandler, C.G., Cornell, J. and Harrison, J. (1996) Nitrogen fertigation requirements of drip-irrigated strawberries. *Journal of the American Society for Horticultural Science* 121, 660–665.

Hochmuth, G.J., Locascio, S., Kostewicz, S.R. and Matin, F.G. (1993) Irrigation method and row cover use for strawberry freeze protection. *Journal of the American Society for Horticultural Science* 118, 575–579.

Hokanson, S. and Finn, C. (1999) Strawberry cultivars for the east. *Fruit Varieties Journal* (in press).

Hokanson, K.E., Harrison, R.E., Luby, J.J. and Hancock, J.F. (1993) Morphological variation in *Fragaria virginiana* from the Rocky Mountains. *Acta Horticulturae* 348, 94–101.

Holevas, C.D. (1966) The effect of vesicular-arbuscular mycorrhiza on the uptake of soil phosphorus by strawberry (*Fragaria* sp. var. Cambridge Favorite). *Journal of Horticultural Science* 41, 57–64.

Honda, F., Matsuda, T., Morishita, M., Iwanaga, Y. and Fushihara, H. (1981) Studies on the breeding of the new strawberry variety 'Terunoka'. *Bulletin of the Vegetable and Ornamental Crops Research Station* 5, 1–14.

Hondelmann, W. (1965) Investigations on breeding for yield in the garden strawberry, *Fragaria ananassa* Duch. (in German). *Pflanzenuchtung* 19, 137–139.

Hortynski, J. (1989) Genotype – environmental interaction in strawberry breeding. *Acta Horticulturae* 265, 175–179.

Howard, C.M. and Albregts, E.E. (1973) A strawberry fruit rot caused by *Dendrophoma obscurans*. *Phytopathology* 63, 419–421.

Howard, C.M., Maas, J.L., Chandler, C.K. and Albregts, E.E. (1992) Anthracnose of strawberry caused by the *Colletotrichum* complex in Florida. *Plant Disease* 76, 976–981.

Howard, C.M., Overman, A.J., Price, J.F. and Albregts, E.E. (1985) *Diseases, Nematodes, Mites and Insects Affecting Strawberries in Florida*. University of Florida Institute of Food and Agriculture, Bulletin 857.

Hsu, C.S., Watkins, R. Bolton, A.T. and Spangelo, L.P.S. (1969) Inheritance of resistance to powdery mildew in the cultivated strawberry. *Canadian Journal of Genetics and Cytology*. 11, 426–438.

Hubbard, N.L., Pharr, D.M. and Huber, S.C. (1991) Sucrose phosphate synthase and other sucrose metabolizing enzymes in fruits of various species. *Physiologia Plantarum* 82, 191–196.

Huber, D.J. (1984) Strawberry fruit softening: the potential roles of polyuronides and hemicelluloses. *Journal of Food Science* 49, 1310–1315.

Huber, D.M., Warren, H.L., Nelson, D.W. and Tsai, C.Y. (1977) Nitrification inhibitors – new tools for food production. *BioScience* 27, 523–529.

Hudson, M.A., Ricketts, V.A. and Holgate, M.E. (1977) Home frozen strawberries. III. Factors affecting sensory assessment. *Journal of Food Technology* 12, 421–426.

Hughes, H.G. and Janick, J. (1974) Production of tetrahaploids in the cultivated strawberry. *HortScience* 9, 422–444.

Hughes, H.M. (1972) Experiments on defoliation of two strawberry cultivars at three centers. *Experimental Horticulture* 3, 50–56.

Hughes, J.D. (1989) Strawberry June yellows – a review. *Plant Pathology* 38, 146–160.

Hughes, M., Martin, L.W. and Breen, P.J. (1978) Mycorrhizal influence on the nutrition of strawberries. *Journal of the American Society for Horticultural Science* 103, 179–181.

Hummel, R.L. and Moore, P.P. (1997) Freeze resistance of Pacific Northwest strawberry flowers. *Journal of the American Society for Horticultural Science* 122, 179–182.

Hummer, K. (1995) What's new in strawberry genetic resources: raw materials for a better berry. In: Pritts, M.P., Chandler, C.K. and Crocker, T.E. (eds) *Proceedings of the IV North American Strawberry Conference*, Orlando, Florida, pp. 79–86.

Hunter, B.G., Richardson, J., Dietzgen, R.G., Karu, A., Sylvester, E.S., Jackson, A.O. and Morris, T.J. (1990) Purification and characterization of strawberry crinkle virus. *Phytopathology* 80, 282–287.

Ichijima, K. (1926) Cytological and genetic studies in *Fragaria*. *Genetica* 11, 590–604.

Ichijima, K. (1930) Studies on the genetics of *Fragaria*. *Zeitschrift für Induktive Abstammungs Vererbungslehre* 55, 300–347.

Irkaeva, R.M. (1993) The genetics of male sterility in the strawberry *Fragaria vesca* L. *Genetica* (Moskova) 29, 1485–1491.

Irkaeva, R.M. and Ankudinova, I.N. (1994) Evaluation of the outbreeding level of the strawberry *Fragaria vesca* L. using genetic markers. *Genetica* 30, 816–822.

Islam, A. S. (1960) Possible role of unreduced gametes in the origin of polyploid *Fragaria*. *Biologia* 6, 189–192.

Ito, H. and Sato, T. (1962) Studies on the flower formation in the strawberry plant. I. Effects of temperature and photoperiod on the flower production. *Tohoku Journal of Agricultural Research* 13, 191–203.

Iwatsubo, Y. and Naruhashi, N. (1989) Karyotypes of three species of *Fragaria* (Rosaceae). *Cytologia* 54, 493–497.

Izhar, S. (1997) Infra short-day strawberry types. *Acta Horticulturae* 439, 155–160.

Izhar, S and Izsak, E. (1995) Infra Short-day Strawberry Plants. United States Patent no. 5, 444, 179.

Izsak, E. and Izhar, S. (1984) Breeding and testing of early strawberry varieties in the central and northern Negere regions. *Hassadeh* 64: 1774–1777. (Hebrew original and English translation.)

Izsak, E. and Izhar, S. (1989) The Israeli strawberry industry. *Acta Horticulturae* 265, 711–715.

James, D.J., Passey, A.J. and Barbara, D.J. (1990) *Agrobacterium*-mediated transformation of the cultivated strawberry (*Fragaria* × *ananassa*) using disarmed binary vectors. *Plant Science* 69, 79–94.

Jamieson, A.R. (1997) New cultivars and selections from the Kentville strawberry breeding program. *Acta Horticulturae* 439, 233–236.

Jamieson, A.R. and Sanford, K.A. (1996) Field performance of June Yellows-affected clones of 'Blomidon' strawberry. *HortScience* 31, 848–850.

Javouhey, M. and Chausset, D. (1997) Marabella: a new short-day strawberry cultivar. *Acta Horticulturae* 439, 269–274.

Jelenkovic, G., Wilson, M.L. and Harding, P.J. (1984) An evaluation of intergeneric hybridization of *Fragaria* spp. × *Potentilla* spp. as a means of haploid production. *Euphytica* 33, 143–152.

Jelkmann, W., Martin, R.R., Lesemann, D. E., Vetten, H.J. and Skelton, F. (1990) A new potexvirus associated with strawberry mild yellow edge disease. *Journal of General Virology* 71, 1251–1258.

Jensen, R.J. and Hancock, J.F. (1981) Multivariate relationships among California strawberries. *Bulletin of the Torrey Botanical Club* 109, 136–147.

Jhooty, J.S. and McKeen, W.E. (1965) Studies on powdery mildew of strawberry caused by *Sphaerotheca macularis*. *Phytopathology* 55, 281–285.

Johanson, F. (1980) *Hunger in Strawberries*. K. and H. Printers, Everett, Washington.

John, M.K., Daubeny, H.A. and McElrow, F.D. (1975) Influence of sampling time on elemental composition of strawberry leaves and petioles. *Journal of the American Society for Horticultural Science* 100, 513–517.

Johnson, H.A. (1990) The contributions of private strawberry breeders. *HortScience* 25, 897-902.

Johnson, J.D. and Ferrell, W.K. (1983) Stomatal response to vapor pressure deficit and the effect of plant water stress. *Plant, Cell and Environment* 6, 451–456.

Jones, D.F. and Singleton, W.R. (1940) The improvement of naturally cross pollinated plants by selection in self fertilized lines. 3. Investigations with vegetatively propagated fruits. Strawberry and raspberry hybrids. *Connecticut Agricultural Experiment Station Bulletin* 435, 325–347.

Jones, J.K. (1955) Cytogenetic studies in the genera *Fragaria* and *Potentilla*. Ph.D dissertation, University of Reading.

Jones, O.P., Waller, B.J. and Beech, M.G. (1988) The production of strawberry plants from callus culture. *Plant Cell, Tissue and Organ Culture* 12, 235–241.

Jones, W.A. and Jackson, C.G. (1990) Mass production of *Anaphes iole* for augmentation against *Lygus hesperus*: effects of fecundity and longevity. *Southwestern Entomologist* 15, 463–468.

Jurik, T.W. (1983) Reproductive effort and CO_2 dynamics of wild strawberry populations. *Ecology* 64, 1329–1342.

Jurik, T.W., Chabot, J.F. and Chabot, B.F. (1979) Ontogeny of photosynthetic performance in *Fragaria virginiana* under changing light regimes. *Plant Physiology* 63, 542–547.

Jurik, T.W., Chabot, J.F. and Chabot, B.F. (1982) Effects of light and nutrients on leaf size, CO_2 exchange, and anatomy in wild strawberry (*Fragaria virginiana*). *Plant Physiology* 70, 1044–1048.

Kaden-Kreuziger, D., Lamprecht, S., Martin, R.R. and Jelkmann, W. (1995) Immunocapture polymerase chain reaction assay and ELISA for the detection of strawberry mild yellow edge associated potexvirus. *Acta Horticulturae* 385, 33–38.

Kader, A.A. (1991) Quality and its maintenance in relation to the postharvest physiology of strawberry. In: Luby, J.J. and Dale, A. (eds) *The Strawberry into the 21st Century*. Timber Press, Portland, Oregon, pp. 145–152.

Kader, A.A. (1994) Modified atmospheres during transport and storage. In: Kader, A. (ed.), *Postharvest Technology of Horticultural Crops*, University of California Publication 3311, pp. 85–92.

Kakouli-Duarte, T, Labuschagne, L. and Hague, N.G.M. (1997) Biological control of the black vine weevil, *Otiorhynchuc sulcatus* (Coleoptera: Curculionidae) with entomopathogenic nematodes (Nematoda: Rhabditida). *Annals of Applied Biology* 131, 11–27.

Kalt, W., Prange, R.K. and Lidster, P.D. (1993) Postharvest color development of strawberries: the influence of maturity, temperature and light. *Canadian Journal of Plant Science* 73, 541–548.

Kaşka, N. (1997) Strawberry growing in Turkey. *Acta Horticulturae* 439, 385–391.

Kaşka, N., Türemiş, N., Kafkas, S. and Çömlekçioğlu, N. (1997) The performance of some strawberry cultivars grown under high tunnels in the climatic condition of Andana (Turkey). *Acta Horticulturae* 439, 291–298.

Katan, J. (1981) Solar heating (solarization) of soil for control of soilborne pests. *Annual Review of Phytopathology* 19, 211–236.

Ke, K., Goldstein, L., O'Mahony, M. and Kader, A.A. (1991) Effects of short-term exposure to low O_2 and high CO_2 atmospheres on quality attributes of strawberries. *Journal of Food Science* 56, 50–54.

Keefer, R.F., Hickman, C.E. and Adams, R.E. (1978) The response of strawberry yields to soil fumigation and nitrogen fertilization. *HortScience* 13, 51–52.

Kerkhoff, K.L., Williams, J.M. and Barden, J.A. (1988) Net photosynthetic rates and growth of strawberry after partal defoliation. *HortScience* 23, 1086.

Khanizadeh, S. and Bélanger, A. (1997) Classification of 92 strawberry genotypes based on their leaf essential oil composition. *Acta Horticulturae* 439, 205–209.

Khanizadeh, S., Hamel, C., Kianmehr, H., Buszard, D. and Smith, D.L. (1995) Effect of three vesicular-arbuscular mycorrhizae species and phosphorus on reproductive and vegetative growth of three strawberry cultivars. *Journal of Plant Nutrition* 18, 1073–1079.

Ki, W.K. and Warmund, M.R. (1992) Low temperature injury to strawberry floral organs at several stages of development. *HortScience* 27, 1302–1304.

Kim, C.H., Cho, W.D. and Kim, S.B. (1982) Studies on varietal resistance and chemical control to the wilt of strawberry caused by *Fusarium oxysporum*. *Korean Journal of Plant Protection* 21, 61–67.

Kim, C.S., Brown, W.G. and Langmo, R.D. (1980) Economic feasibility to Oregon growers of mechanically harvested strawberries (costs). *Agricultural Experiment Station Bulletin* 645, Oregon State University, pp. 175–199.

Kinet, J.M. and Parmentier, A. (1989) Changes in quality of cold-stored strawberry plants (cv. Elsanta) as a function of storage duration: the flowering response in controlled environments. *Acta Horticulturae* 265, 327–334.

King, T.H., Triet, M. and Baskin, A.D. (1950) Sprays to control chlorosis in flax and strawberries grown on alkaline soil in Minnesota. *Phytopathology* 40, 14–15.

Kinnanen, H. and Sako, J. (1979) Irrigation requirements of strawberries. *Ann. Agric. Fenniae* 18, 160–167.

Kirk, D.E. (1980a) Postharvest handling of mechanically harvested strawberries in Oregon. *Agricultural Experiment Station Bulletin* 645, Oregon State University, Portand, pp. 92-102.

Kirk, D.E. (1980b) The concept and design of an on-farm clean-up system for mechanically harvested strawberries. *Agricultural Experiment Station Bulletin* 645, Oregon State University. pp. 126–132.

Kirsch, K. (1959) The importance of interaction effects in fertilizer and lime studies in strawberries. *Proceedings of the American Society for Horticultural Science* 73, 181–188.

Knee, M., Sargent, J.A. and Osborne, D.J. (1977) Cell wall metabolism in developing strawberry fruits. *Journal of Experimental Botany* 8, 377–396.

Kolb, K.A. (1986) Putting together a balanced commercial strawberry fertilizer program. *Proceedings of the North American Strawberry Growers Association Meeting*, pp. 41–44.

Kovach, J., Wilcox, W., Agnello, A. and Pritts, M. (1993) *Strawberry IPM Scouting Procedures*. New York State Integrated Pest Management Program, no. 203b,

Cornell University, Ithaca, New York.

Kronenberg, H.G. and Wassenaar, L.M. (1972) Dormancy and chilling requirement of strawberry varieties for early forcing. *Euphytica* 21, 454–459.

Kronenberg, H.G., Braak, J.P. and Zeilinga, A.E. (1959) Poor fruit in strawberries. II. Malformed fruit in Jucunda. *Euphytica* 8, 245–249.

Kruckeberg, A.R. (1967) Ecotypic response to ultramafic soils by some plant species of north western United States. *Brittonia* 19, 133–151.

Kurppa, S., and Vrain, T.C. (1989) Effects of *Pratylenchus penetrans* on the infection of strawberry roots by *Gnomonia comari. Journal of Nematology* 21, 511–516.

Labarca, E. (1994) *Butamalón*. Editorial Universitaria y Fondo de Cultura Económica. Santiago, Chile.

Lacy, C.N.D. (1973) Phenotypic correlations between vegetative characters and yield components in strawberry. *Euphytica* 22, 546–554.

Lal, S.D. and Seth, J.N. (1979) Studies on genetic variability in strawberry (*Fragaria × ananassa* Duch.). *Progressive Horticulture* 11, 49–53.

Lal, S.D. and Seth, J.N. (1980) Correlation studies in strawberry (*Fragaria × ananassa* Duch.). *Indian Journal of Horticulture* 37, 371–375.

Lal, S.D. and Seth, J.N. (1981) Studies on combining ability in strawberries (*Fragaria × ananassa*) I. Number of inflorescences, number of flowers, days to maturity and number of fruits. *Canadian Journal of Genetics and Cytology* 23, 373–378.

Lal, S.D. and Seth, J.N. (1982) Studies on combining ability in strawberries (*Fragaria × ananassa*) I. Fruit length, fruit diameter, fruit weight, ascorbic acid, total soluble solids and fruit yield. *Canadian Journal of Genetics and Cytology* 24, 479–483.

LaMondia, J.A. and Martin, S.B. (1989) The influence of *Pratylenchus penetrans* and temperature on black root rot of strawberry by binucleate *Rhizoctonia* spp. *Plant Disease* 73, 107–110.

Larsen, M. (1994) Volatile compounds formed in strawberries under anaerobic conditions and their influence on off-flavor formation. *Developmental Food Science* 35, 421–424.

Larsen, M. and Poll, L. (1992) Odour thresholds of some important aroma compounds in strawberries. *Zeitschrift fur Lebensmittel- Untersuchung und-Forschung* 195, 120–123.

Larsen, M. and Watkins, C.B. (1995) Firmness and concentration of aldehyde, ethyl acetate and ethanol in strawberries stored in controlled and modified atmospheres. *Postharvest Biology and Technology* 5, 39–50.

Larson, K.D. (1994) Strawberry. In: Schaffer, B. and Anderson, P.C. (eds) *Handbook of Environmental Physiology of Fruit Crops*, Vol. 1, *Temperate Crops*. CRC Press, Boca Raton, Florida.

Larson, K.D. and Shaw, D.V. (1995a) Strawberry nursery soil fumigation and runner plant production. *HortScience* 30, 236–237.

Larson, K.D. and Shaw, D.V. (1995b) Relative performance of strawberry genotypes in fumigated and non-fumigated soil. *Journal of the American Society for Horticultural Science* 120, 274–277.

Lauro, E.M. (1986) Machinery and equipment. In: *Proceedings of the Strawberry Research Corporation: Annual Report for 1985*, Simcoe, Ontario, pp. 29–40.

Lavín, A. (1997) Caracterización botanica, fisiológica y agronómica de ecotipos chilenos de *Fragania chiloensis* (L.) Duch., recolectados en las X y XI Regiones de Chile. *Informe Final, Proyecto Fondecyt 1940083*.

Lavín, A., Muñoz, C. Ballington, J.R. and Cameron, J.S. (1993) Colección de *Fragaria chiloensis* L.en la X y XI Regiones de Chile. *Simiente* (Chile) 63, 18–20.

Lawrence, F.J. and Martin, L.H. (1980) Breeding strawberries for machine harvest in the pacific northwest. In: Martin, L.W. and Morris, J.R. (eds) *Strawberry Mechanization*. Agricultural Experiment Station Bulletin 645, Oregon State University, Corvallis, pp 30- 37.

Lawrence, F.J., Galletta, G.J. and Scott, D.H. (1990) Strawberry breeding work of the U.S. Department of Agriculture. *HortScience* 25, 895–896.

Lee, B.Y., Takahashi, K. and Sugiyama, T. (1968) Studies on dormancy in strawberry plants. I. Varietal differences in chilling requirement to break dormancy. *Journal of the Japanese Society of Horticulture* 37, 129–134.

Lee, I.M., and Davis, R.E. (1993) Differentiation of strains in the aster yellows mycoplasmalike organism strain cluster by serological assay and monoclonal antibodies. *Plant Disease* 77, 815–817.

Lee, I.M., Davis, R.E., Chen, T.A., Chiykowski, L.N., Fletcher, J., Hiruki, C. and Schaff, D.A. (1992) A genotype-based system for identification and classification of mycoplasmalike organisms (MLOs) in the asters yellows MLO strain cluster. *Phytopathology* 82, 977–986.

Lee, V. (1964) Antoine Nichlos Duchesne – first strawberry hybridist. *American Horticulture Magazine* 43, 80–88.

Le Mière, P., Hadley, P., Darby, J. and Battey, N.H. (1996) The effect of temperature and photoperiod on the rate of flower initiation and the onset of dormancy in the strawberry (*Fragaria × ananassa* Duch.). *Journal of Horticultural Science* 71, 361–371.

Le Mière, P., Hadley, P., Darby, J., and Battey, N.H. (1998) The effect of thermal environment, planting date and crown size on growth, development and yield of *Fragaria × ananassa* Duch. Cv. Elsanta. *Journal of Horticultural Science and Biotechnology* 73, 453–460.

Lenz, F. (1974) Fruit effects on formation and distribution of photosynthetic assimilates. In: Antoszewski, R., Harrison, L. and Nowosielski, J. (eds) *Proceedings of the XIXth International Horticultural Congress*, pp. 155–166.

Leone, G., Linder, J.L. and Schoen, C.D. (1992) Attempts to purify strawberry viruses by nonconventional separation methods. *Acta Horticulturae* 308, 121–129.

Lesourd, F. (1949). *Le Fraisier*. Maison Rustique, Paris.

Levy, A., Rowland, L.J., Galletta, G.J., Martelli, G. and Greco, I. (1994) Identification of strawberry genotypes and evaluation of their genetic relationships using randomly amplified polymorphic DNA (RAPD) analysis. *Advances in Strawberry Research* 13, 36–39.

Levitt, J. (1980) *Responses of Plants to Environmental Stresses*, Vol. I. *Chilling, Freezing and High Temperature Stresses*, 2nd edn. Academic Press, New York.

Leyel, H. (1926) *The Magic of Herbs, a Modern Book of Secrets.* Harcourt, Brace, New York.

Li, C. and Kader, A.A. (1989) Residual effects of controlled atmospheres on postharvest physiology and quality of strawberries. *Journal of the American Society of Horticultural Science* 114, 629–634.

Lieten, F. (1993) Methods and strategies of strawberry forcing in Europe: historical perspectives and recent developments. *Acta Horticulturae* 348, 158–170.

Lieten, F. (1997) Effect of CO_2 enrichment on greenhouse grown strawberry. *Acta Horticulturae* 439, 583–588.

Lieten, F. and Baets, W. (1991) Greenhouse strawberry culture in peat bags. *Advances in Strawberry Production* 10, 56–57.

Lieten, F. and Goffings, G. (1997) Effect of temperature and controlled atmosphere on cold storage of strawberry plants. *Acta Horticulturae* 439, 445–448.

Lieten, F. and Marcelle, R.D. (1993) Relationships between fruit mineral content and the 'albinism' disorder in strawberry. *Annals of Applied Botany* 123, 433–439.

Lilienfeld, F.A. (1933) Cytological and genetical studies in *Fragaria*. I. Fertile tetraploid hybrid between *F. nipponica* (n=7) and *F. elatior* (n=21). *Japananese Journal of Botany* 6, 425-458.

Lindeman, J. and Suslow, T.V. (1987) Competition between ice nucleation-active wild type and ice nucleation-deficient deletion mutant strains of *Pseudomonas syringae* and *P. fluorescens* biovar 1 and biological control of frost injury on strawberry blossoms. *Phytopathology* 77, 882–886.

Liu, Z.R. and Sanford, J.C. (1988) Plant regeneration by organogenesis from strawberry leaf and runner tissue. *HortScience* 23, 1057–1059.

Locascio, S. J. (1980) Nutrition application through trickle systems. In: Childers, N.F. (ed.) *The Strawberry: Cultivars to Marketing*. Horticultural Publications, Gainesville, Florida, pp. 95–99.

Locascio, S.J., and Martin, F.G. (1985) Nitrogen source and application timing for trickle irrigation. *Journal of the American Horticulture Society* 110, 820–823.

Lockhart, C.L. (1967) Effect of temperature and various CO_2 and O_2 concentrations on growth of *Typhula* sp., a parasitic fungus of strawberry plants. *Canadian Journal of Plant Science* 47, 450–452.

Lockhart, C.L. and Eaves, C.A. (1966). The influence of controlled atmospheres on the storage of strawberry plants. *Canadian Journal of Plant Science* 46, 151–154.

Long, J.H. (1935) Seasonal changes in nitrogen and carbohydate content of the strawberry plant. *Proceedings of the American Society for Horticultural Science* 33, 386–388.

López-Aranda, J.M. and Bartual, R. (1999) Strawberry production in Spain. Cost Action 836: W.G.2 and M.C. Meetings. Malaga, Spain, March 4–6.

López-Galarza, S., Maroto, J.V., Pascual, B., Bono, M.S. and Alagarda, J. (1993) Influence of different climatic protection and forcing systems on some production parameters of strawberry (*Fragaria* × *ananassa* Duch) in Spain. *Acta Horticulturae* 348, 249–251.

López-Serrano, M. and Ros Barceló, A. (1996) Purification and characterization of a basic peroxidase isoenzyme from strawberries. *Food Chemistry* 55, 133–137.

Lou, Y. and Liu, X. (1991) Effects of ethylene on RNA metabolism in strawberry fruit after harvest. *Journal of Horticultural Science* 69, 137–139.

Luby, J. J. and Stahler, M.M. (1993) Collection and evaluation of *Fragaria virginiana* in North America. *Acta Horticulturae* 345, 49–54.

Luby, J.J., Hancock, J.F. and Ballington, J.R. (1992) Collection of native strawberry (*Fragaria* ssp.) germplasm in the Pacific Northwest and Northern Rocky Mountains of the USA. *HortScience* 27, 12–17.

Luby, J. J., Hancock, J.F. and Cameron, J.S (1991) Expansion of the strawberry germplasm base in North America. In: Dale, A. and Luby, J.J. (eds) *The Strawberry into the 21st Century*. Timber Press, Portland, Oregon, pp. 66–75.

Luby, J., Harrison, R., Furnier, G. and Hancock, J. (1995) Germplasm resources for strawberries: morphological traits. In: Pritts, M.P., Chandler, C.K. and Crocker,

T.E. (eds) *Proceedings of the IV North American Strawberry Conference*, Orlando, Florida, pp. 95–102.

Lucas, E.R. (1979) Cryogenic pre-cooling/air cushion freezing reduce damage, moisture loss to nil. *Quick Frozen Foods* 42, 26–33.

Luffman, M. and Macdonald, P.J. (1993) Fragaria germplasm at the Canadian Clonal Genebank. *Acta Horticulturae* 348, 102–107.

Lundahl, D.S., McDaniel, M.R. and Wrolstad, R.E. (1989) Flavor, aroma, and compositional changes in strawberry juice stored at 20°C. *Journal of Food Science* 54, 1255–1258.

Lundergan, C.A. and Moore, J.N. (1975) Inheritance of ascorbic acid content and color intensity in fruits of strawberry (*Fragaria* × *ananassa* Duch.). *Journal of the American Society for Horticultural Science* 100, 633–635.

Maas, J.L. (1985) New symptoms of strawberry leaf blight disease. *Advances in Strawberry Production* 4, 34–35.

Maas, J.L. (ed.) (1998) *Compendium of strawberry diseases.* American Phytopathological Society, St Paul, Minnesota.

Maas, J.L. and Galletta, G.J. (1989) Germplasm evaluation for resistance to fungus-incited diseases. *Acta Horticulturae* 265, 461–472.

Maas, J.L. and Galletta, G.L. (1997) Recent progress in strawberry disease research. *Acta Horticulturae* 439, 769–779.

Maas, J.L. and Smith, W.L. (1978) 'Earliglow' a possible source of resistance to *Botrytis* fruit rot in strawberry. *HortScience* 13, 275–276.

Maas, J.L., Harker, J.R. and Galletta, G.J. (1988) Occurence of an exotic race of *Phytophthora fragarie* in Maine. *Advances in Strawberry Production* 8, 42–44.

Maas, J.L., Galletta, G.J. and Draper, A.D. (1989) Resistance in strawberry to races of *Phytophthora fragariae* and to isolates of *Verticillium* from North America. *Acta Horticulturae* 265, 521–526.

Maas, J.L., Galletta, G.J. and Stoner, G.D. (1991a) Ellagic acid, and anticarcinogen in fruits especially strawberries: a review. *HortScience* 26, 10–14.

Maas, J.L., Wang, S.Y. and Galletta, G.J. (1991b) Evaluation of strawberry cultivars for ellagic acid content. *HortScience* 26, 66–68.

Maas, J.L., Pooler, M.R. and Galletta, G.J. (1995) Bacterial angular leaf spot disease of strawberry: present status and prospects for control. *Advances in Strawberry Research* 14, 18–24.

Maas, J.L., Wang, S.Y. and Galletta, G.J. (1996) Health enhancing properties of strawberry fruit. In: Pritts, M.P., Chandler C.K. and Crocker, T.E. (eds) *Proceedings of the IV North American Strawberry Conference*, Orlando, Florida, pp. 11–18.

Maas, J.L., Gouin-Behe, Hartung, J.S. and Hokanson, S.C. (1999) Putative sources in strawberry to bacterial angular leafspot. *HortScience* 34, 457.

MacFarlane Smith, W.H. and Jones, J.K. (1985) Intergeneric crosses with *Fragaria* and *Potentilla*. II. Crosses between the progeny of *Fragaria moschata* × *Potentilla fruticosa* and the original parents. *Euphytica* 34, 737–744.

MacFarlane Smith, W. H., Jones, J.K. and Sebastiampillai, A.R. (1989) Pollen storage of *Fragaria* and *Potentilla*. *Euphytica* 41, 65–69.

MacIntyre, D. and Gooding, H.J. (1978) The assessment of strawberries for decapping by machine. *Horticultural Research* 18, 127–136.

MacLachlan, J.B. (1974) The inheritance of color of fruit and the assessment of plants as sources of color in the cultivated strawberry. *Horticultural Research* 14, 29–39.

Macoun, W.T. (1924) *Report of the Dominion Horticulturalist*. Ottawa, Canada.

Madden, L.V. and Ellis, M.A. (1988) How to develop plant disease forecasters. In: Kranz, J. and Roten, J. (eds) *Techniques in Epidemiology*. Springer-Verlag, Berlin.

Madden, L.V. and Ellis, M.A. (1990) Effect of ground cover on splash dispersal of *Phytophthora cactorum* from strawberry fruits. *Journal of Phytopathology* 129, 170–174.

Madden, L.V., Ellis, M.A., Grove, G.G., Reynolds, K.M. and Wilson, L.L. (1991) Epidemiology and control of leather rot of strawberries. *Plant Disease* 74, 439–446.

Madden, L.V., Wilson, L.L. and Ellis, M.A. (1993) Field spread of anthracnose fruit rot of strawberry in relation to ground cover and ambient weather conditions. *Plant Disease* 77, 861–866.

Mailloux, G. and Bostanian, N.J. (1993) Development of the strawberry bud weevil in strawberry fields. *Annals of the Entomological Society of America* 86, 384–393.

Mangelsdorf, A.J. and East, E.M. (1927) Studies on the genetics of *Fragaria*. *Genetics* 12, 307–339.

Manning, K. (1993) Soft fruits. In: Seymore, G.B., Taylor, J.E. and Tucker, G.E. (eds) *Biochemistry of Fruit Ripening*. Chapman & Hall, London, pp. 347–378.

Manning, K. (1994) Changes in gene expression during strawberry ripening and their regulation by auxin. *Planta* 194, 62–68.

Manning, K. (1997) Ripening enhanced genes of strawberry: their expression, regulation and function. *Acta Horticulturae* 439, 165–167

Manning, K. (1998) Genes for fruit quality in strawberry. In: Cockshull, K.E., Gray, D., Seymour, G.B. and Thomas, B (eds) *Genetic and Environmental Manipulation of Horticultural Crops*. CAB International, Wallingford, UK, pp. 51–61.

Mannini, P. and Anconelli, S. (1993) Leaf temperature and water stress. *Acta Horticulturae*. 345, 55–61.

Marini, R. P. and Boyce, B.R. (1979) Influence of low temperatures during dormancy on growth and development of 'Catskill' strawberry plants. *Journal of the American Society for Horticultural Science* 104, 159–162.

Marschner, H. (1986) *Mineral Nutrition of Higher Plants*. Academic Press, London.

Martin, R.R. (1995) New directions in research on viral pathogens of strawberry. In: Pritts, M.P., Chandler, C.K. and Crocker, T.E. (eds) *Proceedings of the IV North American Strawberry Conference*, University of Florida, Orlando, pp. 48–53.

Martinsson, M. (1997) Strawberry growing in Sweden. *Acta Horticulturae* 439, 397-403.

Mason, D.T. (1966) Inflorescence initiation in the strawberry I. Initiation in the field and its modification by post-harvest defoliation. *Horticultural Research* 6, 33–44.

Mason, D.T. and Rath, R. (1980). The relative importance of some yield components in East of Scotland strawberry plantations. *Annals of Applied Biology* 95, 399–408.

Mathews, H., Wagoner, W., Kellogg, J. and Bestwick, R. (1995) Genetic transformation of strawberry: stable integration of a gene to control the biosynthesis of ethylene. *In Vitro Cellular and Developmental Biology* 31, 36–43.

Maxie, E.C. and Abdel-Kader, A. (1966) Food irradiation – physiology of fruits as related to the feasibility of the technology. *Advances in Food Research* 15, 105–138.

Maxie, E.C., Mitchell, F.G. and Greathead, A. (1959) Studies on strawberry quality. *California Agriculture* 13, 11–16.

May, G. and Pritts, M.P. (1990a) Strawberry nutrition. *Advances in Strawberry Production* 9, 10–24.

May, G. and Pritts, M.P. (1990b) Interactions among phosphorous, boron, and zinc in 'Earliglow' strawberry. *Proceedings North American Strawberry Growers Association*, pp. 39–54.

May, G.M. and Pritts, M.P. (1994) Seasonal patterns of growth and tissue nutrient content in strawberries. *Journal of Plant Nutrition* 17, 1149–1162.

Maynard, D.N., Hochmuth, G.J., Vavrina, C.S., Stall, W.M., Kucharek, T.A., Johnson, F.A. and Taylor, T.G. (1996) Strawberry production in Florida. In: *Vegetable Production Guide for Florida*, Florida Cooperative Extension Service, University of Florida, SP 170.

McCulloch, J. (1978) Strawberry crimp. *Queensland Agricultural Journal*, 104, 345–347.

McDonald, S.S., Archbold, D.D. and Lailiang, C. (1994) Evaluation of g_{ti} method to assess heat and desiccation injury among strawberry species. *HortScience* 29, 477.

McElhannon, W.S. and Mills, H.A. (1981) Suppression of denitrification with nitrapyrin. *HortScience* 16, 530–531.

McGrew, J.R. (1980) Meristem culture for production of virus-free strawberries. In: *Proceedings of the Conference on Nursery Production of Fruit Plants Through Tissue Culture – Applications and Feasibility*, Agriculture Research Results, ARR-NE–11, pp. 80–85.

McNiesh, S.S., Welch, N.C. and Nelson, R.D. (1985) Trickle irrigation requirements for strawberries in coastal California. *Journal of the American Society for Horticultural Science* 110, 174–178.

McNicol, R.J., Graham, J. and Kerby, N.W. (1997) Recent advances in strawberry breeding and product development at SCRI. *Acta Horticulturae* 439, 129–132.

Medina-Escobar, N., Cárdenas, J., Muñoz-Blanco, J. and Caballero, J.L. (1998) Cloning and molecular characterization of a strawberry ripening-related cDNA corresponding to a mRNA for low-molecular-weight-heat-shock protein. *Plant Molecular Biology* 36, 33–42.

Meesters, P. and Pitsioudis, A. (1997) Cultivating the day-neutral cultivar Selva under Belgian conditions. *Acta Horticulturae* 439, 407- 414.

Melville, A.H., Draper, A.D. and Galletta, G.J. (1980a) Transmission of red stele resistance by inbred strawberry selections. *Journal of the American Society for Horticultural Science* 105, 608–610.

Melville, A.H., Galletta, G.J., Draper, A.D. and Ng, T.J. (1980b) Seed germination and early seedling vigor in progenies of inbred strawberry selections. *HortScience* 15, 49–750.

Milholland, R.D. (1996) A monograph of *Phytophthora fragariae* and the red stele disease of strawberry. *North Carolina Research Service Technical Bulletin* 306.

Milholland, R.D. and Daykin, M.E. (1993) Colonization of roots of strawberry cultivars with different levels of susceptibility to *Phytophthora fragariae*. *Phytopathology* 83, 538–542.

Milholland, R.D., Ritchie, D.F., Daykin, M.E. and Gutierrez, W.A. (1996) Multiplication and translocation of *Xanthomonas fragariae* in strawberry. *Advances in Strawberry Research* 15, 13–17.

Miller, A.R. and Chandler, C.K. (1990) Plant regeneration from excised cotyledons of mature strawberry achenes. *HortScience* 25, 569–571.

Miller, P.W. (1959) An improved method of testing the tolerance of strawberry varieties and new selected seedlings to virus infection. *Plant Disease Reporter* 43, 1247–1249.

Miller, P.W. and Waldo, G.F. (1959) The virus tolerance of *Fragaria chiloensis* compared with the Marshall variety. *Plant Disease Reporter* 43, 1120–1131.

Miller, W.R., Davis, P.L. Dow, A. and Bongers, A.J. (1983) Quality of strawberries packed in different consumer units and stored under simulated air-freight shipping conditions. *HortScience* 18, 310–312.

Mills, H.A. and Pokorny, F.A. (1985) Effectiveness of multiple applications of nitapyrin in increasing nitrogen utilization in an organic medium. *HortScience* 20, 869–870.

Misic, P.D. (1965) Occurrence of June yellows and symptoms of white streaks among strawberry hybrid seedlings. *Journal of Scientific Agricultural Research Yugoslavia* 18, 152–159.

Miszczak, A., Forney, C.F. and Prange, R.K. (1995) Development of aroma volatiles and color during postharvest ripening of 'Kent' strawberries. *Journal of the American Society for Horticultural Science* 120, 650–655.

Mitchell, F.G. (1994a) Cooling methods. In: Kader, A.A. (ed.) *Postharvest Technology of Horticultural Crops.* University of California Publication 3311, pp. 56–62.

Mitchell, F.G. (1994b) Postharvest handling system: small fruits (table grapes, strawberries, kiwifruit). In: Kader, A.A. (ed.) *Postharvest Technology of Horticultural Crops.* University of California Publication 3311, pp. 223–240.

Mitchell, F.G., Maxie, E.C. and Greathead, A.S. (1964) Handling strawberries for fresh market. *California Agriculture Experiment Station Circular* 527.

Mitchell, W.C. and Jelenkovic, G. (1995) Characterizing NAD- and NADP-dependent alcohol dehyrogenase enzymes of strawberries. *Journal of the American Society for Horticultural Science* 120, 798–801.

Miura, H., Slhimizu, A. and Imada, S. (1993) Sensitive stage of strawberry fruit to light for coloration. *Acta Horticulturae* 345, 63–65.

Miura, H., Yoshida, M and Yamasaki, A. (1994) Effect of temperature on the size of strawberry fruit. *Journal of the Japanese Society of Horticultural Science* 62, 769–774.

Mochizuki, T. (1995) Past and present strawberry breeding programs in Japan. *Advances in Strawberry Research* 14, 9–17.

Mochizuki, T., Noguchi, Y., Sone, K. and Morishita, M. (1997) Aroma components of amphiploid strawberries derived from interspecific hybrids of *Fragaria* ×*ananassa* and diploid wild species. *Acta Horticulturae* 439, 75–80.

Mohan, S.B., Kennedy, D.M. and Duncan, J.M. (1989) Assessment of red core resistance of strawberry using enzyme-linked immunosorbant assay (ELISA). *Journal of Phytopathology* 126, 97–103.

Mokkila, M., Randell, K., Sariola, J., Hägg, M. and Häkkinen, U. (1997) Improvement of the postharvest quality of strawberry. *Acta Horticulturae* 439, 553–557.

Montgomerie, I. (1967) Pathogenicity of British isolates of *Phytophthora fragariae* and their relationship with American and Canadian races. *British Transactions of the Mycological Society* 50, 57–67.

Moon, B.J., Chung, H.S. and Park, H.C. (1995) Studies on antagonism of *Trichoderma* species to *Fusarium oxysporum* f.sp. *fragariae* V. Biological control of Fusarium wilt of strawberries by a mycoparasite, *Trichoderma harzianum. Korean Journal of Plant Pathology* 11, 298–303.

Moon, J.W., Bailey, D.A., Fallahi, E., Jensen, R.G. and Zhu, G. (1990) Effect of nitrogen application on growth and photosynthetic nitrogen use efficiency in two ecotypes of wild strawberry, *Fragaria chiloensis. Physiologia Plantarum* 80, 612–618.

Moore, J.N. (1968) Effects of post-harvest defoliation on strawberry yields and fruit size. *HortScience* 3, 45–46.

Moore, J.N., Scott, D.H. and Converse, R.H. (1964) Pathogenicity of *Phytophthora fragariae* to certain *Potentilla* species. *Phytopathology* 54, 173–176.

Moore, J.N., Brown, G.R. and Brown, E.D. (1970) Comparison of factors influencing fruit size in large-fruited and small-fruited clones of strawberry. *Journal of the American Society for Horticultural Science* 95, 827–831.

Moore, J.N., Brown, G.R. and Bowen, H.L. (1975) Evaluation of strawberry clones for adaptability to once-over mechanical harvest. *HortScience* 10, 407–408.

Morita, H. (1968) Physiologic races of *Phyophthora fragariae. Shizuoka Agricultural Experiment Station Bulletin 13*.

Morris, J.R. (1980) Strawberry mechanization: a total system approach. In: Childers, N.F. (ed.) *The Strawberry: Cultivars to Marketing.* Horticultural Publications, Gainesville, Florida, pp. 354–372.

Morris, J.R. and Cawthon, D.L. (1980) Postharvest holding of machine-harvested strawberries. In: Childers, N.F. (ed.) *The Strawberry: Cultivars to Marketing.* Horticultural Publications, Gainesville, Florida, pp. 385–413.

Morris, J.R., Kattan, A.A., Nelson, G.S., and Cawthon, D.L. (1978) Developing a mechanized system for production, harvesting and handling of strawberries. *HortScience* 13, 413–422.

Morris, J.R., Sistrunk, W.A., Kattan, A.A., Nelson, G.S. and Cawthon, D.L. (1979a) Yield, quality and utilization of mechanically harvested strawberries for processing. *Food Technology* 35, 92–94.

Morris, J.R., Spayd, S.E., Cawthon, D.L., Kattan, A.A. and Nelson, G.S. (1979b) Strawberry clonal fruit yield and quality responses to hand picking prior to once-over machine harvest. *Journal of the American Society of Horticultural Science* 104, 864–867.

Morris, J.R., Sistrunk, W.A., Nelson, G.S., Spayd, S.E. and Cawthon, D.L. (1980) Quality of mechanically harvested strawberries for processing. *Agricultural Experiment Station Bulletin* 645, Oregon State University, Portland, pp. 144–149.

Morris, J.R., Sims, C.A. and Cawthon, D.L. (1985) Effects of production systems, plant populations and harvest dates on yield and quality of machine-harvested strawberries. *Journal of the American Society for Horticultural Science* 110, 718–721.

Morris, J.R., Sistrunk, W.A., Sims, C.A., Main, G.L. and Wehunt, E.J. (1985b) Effects of cultivar, postharvest storage, preprocessing dip treatments and style of pack on the processing quality of strawberries. *Journal of the American Society of Horticultural Science* 110, 172–177.

Morris, J.R., Main, G.L. and Sistrunk, W.A. (1991) Relationship of treatment of fresh strawberries to the quality of frozen fruit and preserves. *Journal of Food Quality* 14, 467–479.

Morrow, E.B. and Darrow, G.M. (1941) Inheritance of some characteristics in strawberry varieties. *Proceedings of the American Society for Horticultural Science* 59, 269–276.

Morrow, E.B. and Darrow, G.M. (1952) Effects of limited inbreeding in strawberries. *Proceedings of the American Society for Horticultural Science* 59, 269–276.

Morrow, E.B., Comstock, R.E. and Kelleher, T. (1958) Genetic variances in strawberry varieties. *Proceedings of the American Society for Horticultural Science* 72, 170–185.

Moyls, A.L., Sholberg, P.L. and Gaunce, A.P. (1996) Modified atmosphere packaging of grapes and strawberries fumigated with acetic acid. *HortScience* 31, 414–416.

Mudge, K.M., Narayanan, K.R. and Poovaiah, B.W. (1981) Control of strawberry fruit set and development with auxins. *Journal of the American Society for Horticultural Science* 106, 80–84.

Mullen, R.H. and Schlegel, D.E. (1976) Cold storage maintenance of strawberry meristem plantlets. *HortScience* 11, 100–101.

Murant, A.F. and Lister, R.E (1987) European nepoviruses in strawberry. In: Converse, R.H. (ed.)*Virus Diseases of Small Fruits*. USDA/ARS, Washington, DC, pp. 46–52.

Murashige, T. and Skoog, F. (1962) A revised medium for rapid growth and bioassays with tobacco tissue cultures. *Physiologia Plantarum* 15, 473–497.

Murawski, H. (1968) Research on the heritability among strawberry varieties for height of flower stems, powdery mildew resistance, fruit color, flesh color, and berry shape (in German). *Archiev für Gartenbenbau* 16, 293–318.

National Academy of Sciences. (1989) Water soluble vitamins. In: *Recommended Dietary Allowances*, 10th Edn, National Academy Press, Washington, DC, pp. 115–126.

Naumann, W. -D. (1961) Die Wirkung zeitlich begrenzter Wassergaben auf Wuchs- und Ertragleistung von Erdbeeren. *Gartenbau* 26, 441–458.

NDong, C., Quellet, F., Houde, M. and Sarhan, F. (1997) Gene expression during cold acclimation in strawberry. *Plant Cell Physiology* 38, 863–870.

Nehra, N.S., Chibbar, R.N., Hartha, K.K., Dalta, R.S.S., Crosby, W.L. and Stushnoff, C. (1990) Agrobacterium-mediated transformation of strawberry calli and recovery of transgenic plants. *Plant Cell Reports* 9, 10–13.

Nehra, N.S., Stushnoff, C. and Kartha, K.K. (1989) Direct shoot regeneration from strawberry leaf discs. *Journal of the American Society for Horticultural Science* 114, 1014–1018.

Neilson, B.V. and Eaton, G.W. (1983) Effects of boron nutrition upon strawberry yield components. *HortScience* 18, 932–934.

Nemec, S. (1971) Studies on resistance of strawberry varieties and selections to *Mycosphaerella fragariae* in southern Illinois. *Plant Disease Reporter* 55, 573–576.

Nemec, S. and Blake, R.C. (1971) Reaction of strawberry cultivars and their progenies to leaf scorch in southern Illinois. *HortScience* 6, 497–498.

Nes, A. (1997) Evaluation of strawberry cultivars in Norway. *Acta Horticulturae* 439, 275–279.

Nestby, R. (1985) Effect of planting date and defoliation on three strawberry cultivars. *Acta Agricultural Scandinavia* 35, 206–212.

Newenhouse, A.C. and Dana, M.N. (1989) Grass living mulch for strawberries. *Journal of the American Society for Horticultural Science* 114, 859–862.

Nicoll, M. F. and Galletta, G.J. (1987) Variation in growth and flowering habits of Junebearing and everbearing strawberries. *Journal of the American Society for Horticultural Science* 112, 872–880.

Niemurowicz-Szczytt, K. and Zakrzewska, Z. (1981) *Fragaria × ananassa* anther culture. *Bulletin de l'Academie Polonaise des Sciences* Ch. II. Vol. XXVII (5), 341–347.

Nishi, S. and Oosawa, K. (1973) Mass propagation method of virus free strawberry plants through meristem callus. *Japanese Agriculture Research Quarterly* 7, 189–194.

Nishizawa, T. (1990) Effects of daylength on cell length and cell number in strawberry petioles. *Journal of the Japanese Society of Horticultural Science* 59, 533–539.

Nishizawa, T. (1994) Comparison of carbohydrate partitioning patterns between fruiting and defoliated June-bearing strawberry plants. *Journal of the Japanese Society of Horticultural Science* 62, 795–800.

Nishizawa, T. and Hori, Y. (1988) Photosynthesis and translocation of recently assimilated carbon in vegetative and dormant stages of strawberry plants. *Journal of the Japanese Society of Horticultural Science* 57, 633–641.

Nishizawa, T. and Hori, Y. (1993) Effects of defoliation and root heating during rest on leaf growth in strawberry plants. *Tohoku Journal of Agricultural Research* 43, 3–4.

Nitsch, J. P. (1950) Growth and morphogenesis of the strawberry as related to auxin. *American Journal of Botany* 37, 211–215.

Nogata, Y., Ohta, H. and Voragen, A.G.J. (1993) Polygalacturonase in strawberry fruit. *Phytochemistry* 34, 617–620.

Noguchi, Y., Mochizuki, T. and Sone, K. (1997) Interspecific hybrids originated from crossing Asian wild strawberries (*Fragaria nilgerrensis* and *F. iinumae*) to *F. × ananassa*. *HortScience* 32, 439.

Noling, J.W. and Becker, J.O. (1994) The challenge of research and extension to define and implement alternatives to methyl bromide. *Journal of Nematology* 26 (Suppl.), 573–586.

Nordmeyer, D. (1992) The search for novel nematicidal compounds. In: Gommers, F.J. and Maas, P.W.T. (eds) *Nematology from Molecule to Ecosystem*. European Society of Nematologists, Invergowrie, Dundee, Scotland, pp. 281–293.

Norman, J.R., Atkinson, D. and Hooker, J.E. (1996) Arbuscular mycorrhizal fungal-induced alteration to root architecture in strawberry and induced resistance to the root pathogen *Phytophthora fragariae*. *Plant and Soil* 185, 191–198.

Norton, A.P. and Welter, S.C. (1995) Parasitoid releases for *Lygus* management in strawberry. In: Pritts, M.P., Chandler, C.K. and Crocker, T.E. (eds) *Proceedings of the IV North American Strawberry Conference*, University of Florida, Orlando, pp. 59–65.

Norton, A.P. and Welter, S.C. (1996) Augmentation of the egg parasitoid *Anaphes iole* (Hymenoptera: Mymaridae) for *Lygus hesperus* (Heteroptera:Miridae) management in strawberries. *Environmental Entomology* 25, 1406–1414.

Norton, A.P., Welter, S.C., Flexner, J.L., Jackson, C.G., Debolt, J.W. and Pickel, C. (1992) Parasitism of *Lygus hesperus* (Miridae) by *Anaphes iole* (Mymaridae) and *Leiophron uniformis* (Braconidae) in California strawberry. *Biological Control* 2, 131–137.

Nunes, M.C.N, Brecht, J.K., Morais, A.M.B. and Sargent, S.A. (1995) Physical and chemical quality characteristics of strawberries are reduced by a short delay to cooling. *Postharvest Biology and Technology* 6, 17–28.

Nylund, R.E. (1950) The use of 2, 4-D for the control of weeds in strawberry plantings. *Proceedings of the American Society for Horticultural Sciences* 55, 271–275.

Nyman, M. and Wallin, A. (1992a) Improved culture technique for strawberry (*Fragaria × ananassa*) protoplasts and the determination of DNA content in protoplast derived plants. *Plant Cell, Tissue, and Organ Culture* 30, 127–133.

Nyman, M and Wallin, A. (1992b) Transient gene expression in strawberry (*Fragaria ×
ananassa* Duch.) protoplasts and the recovery of transgenic plants. *Plant Cell
Reports* 11, 105–108.

Obrycki, J.J., Ormord, A.M., Gabriel, A.D. and Orr, C.J. (1993) Larval and pupal
parasitism of the strawberry leafroller (Lepidoptera: Tortricidae). *Environmental
Entomology* 22, 679–683.

Oda, Y. (1991) The strawberry in Japan. In: Dale, A. and Luby, J.J. (eds) *The Strawberry
into the 21st Century*. Timber Press, Portland, Oregon, pp. 36–46.

Oda, Y. (1997) Effects of light intensity, CO_2 concentration and leaf temperature on
gas exchange of strawberries – feasibility studies on CO_2 enrichment in Japanese
conditions. *Acta Horticulturae* 439, 403–406.

Oda, Y. and Kawata, K. (1993) Cement block wall cultivation of strawberry in Japan.
Acta Horticulturae 348, 219–226.

Ohwi, J. (1965) *Flora of Japan*. Smithsonian Institution, Washington, DC.

O'Neill, S.D. (1983) Role of osmotic potential gradients during water stress and leaf
senescence in *Fragaria virginiana*. *Plant Physiology* 72, 931–937.

Otterbacher, A.G. and Skirvin, R.M. (1978) Derivation of the binomial *Fragaria ×
ananassa* for the cultivated strawberry. *HortScience* 13, 637–639.

Ourecky, D.K. and Bourne, M.C. (1968) Measurement of strawberry texture with an
Instron machine. *Proceedings of the American Society for Horticultural Science* 93,
317–325.

Ourecky, D.K. and Reich, J.E. (1976) Frost tolerance in strawberry cultivars.
HortScience 11, 413–414.

Ourecky, D.K. and Slate, G.L. (1967) Behavior of the everbearing characteristics in
strawberry. *Proceedings of the American Society for Horticultural Science* 91,
236–241.

Overcash, J.P., Fister, L.A. and Drain, B.D. (1943) Strawberry breeding and the
inheritance of certain characteristics. *Proceedings of the American Society for
Horticultural Science* 42, 435–440.

Overholser, E.L. and Claypool, L.L. (1931) The relation of fertilizers to respiration and
certain physical properties of strawberries. *Proceedings of the American Society for
Horticultural Science* 28, 220–224.

Owen, H.R. and Miller, A.R. (1996) Haploid plant regeneration from anther cultures
of three North American cultivars of strawberry (*Fragaria × ananassa*). *Plant Cell
Reports* 15, 905–909.

Oydvin, J. (1980) Observations on two-spotted mite, *Tetranychus urticae* Koch.
Strawberry mite, *Steneotarsonemus pallidusa* Banks, and strawberry mildew,
Sphaerotheca macularis (Wallr.) Magn. *Forskning og Forsoek 1 Landbruket* 31, 1–9.

Parent, J.G. and Pagé, D. (1995) Authentication of the 13 strawberry cultivars of
Quebec's certification program by random amplified polymorphic DNA analysis
(RAPD). *Canadian Journal of Plant Science* 75, 221–224.

Parry, S. and Ramsdall, D. (1983) *Strawberry Diseases in Michigan*. Extension Bulletin
E-1728, Cooperative Extension Service, Michigan State University, East Lansing.

Patel, A.J. and Sharma, G.S. (1977) Nitrogen release characteristics of controlled
release fertilizer during a four month soil incubation. *Journal of the American
Society for Horticultural Science* 102, 364–367.

Paquin, R., Bolduc, R., Zizka, J., Pelletier, G. and Lechausser, P. (1989) Tolérance au gel
et teneur en sucres et en proline du colet du fraisier (*Fragaria × ananassa* Duch)

durant l'hiver. *Canadian Journal of Plant Science* 69, 945–954.

Pearl, R.T. (1928) *The History of the Cultivated Strawberry*. Southeastern Agriculture College, Wye, Kent.

Pelofske, P.J. and Lawrence, F.J. (1984) Inheritance of size relationship of primary and secondary berries of strawberry. *HortScience* 19, 641–642.

Pelofske, P.J. and Martin, L.W. (1982) Effects of mechanical harvest and defoliation on subsequent performance of three strawberry cultivars. *HortScience* 17, 211–212.

Peng, G. and Sutton, J.C. (1991) Evaluation of microorganisms for biocontrol of *Botrytis cinerea* in strawberry. *Canadian Journal of Plant Pathology* 13, 247–257.

Peng, G., Sutton, J.C. and Kevan, P.G. (1992) Effectiveness of honey bees for applying the biological control agent *Gliocladium roseum* to strawberry flowers to suppress *Botrytis cinerea*. *Canadian Journal of Plant Pathology* 14, 117–129.

Pepin, H.S. and Daubeny, H.A. (1964) The relationship of English and American races of *Phytophthora fragariae*. *Phytopathology* 54, 241.

Pepin, H.S. and Daubeny, H.A. (1966) Reaction of strawberry cultivars and clones of *Fragaria chiloensis* to six races of *Phytophthora fragariae*. *Phytopathology* 56, 361–362.

Pérez, A.G., Olías, R., Sanz, C. and Olías, J.M. (1996) Furanones in strawberries: evolution during ripening and postharvest shelf life. *Journal of Agricultural Food Chemistry* 44, 3620–3624.

Pérez, A.G., Ríos, J.J., Sanz, C. and Olías, J.M. (1992) Aroma components and free amino acids in strawberry variety Chandler during ripening. *Journal of Agricultural Food Chemistry*. 40, 2232–2235.

Pérez, A.G., Ríos, J.J., Sanz, C. and Olías, J.M. (1993) Partial purification and some properties of alcohol acyltransferase from strawberry fruit. *Journal of Agricultural Food Chemistry* 41, 1462–1466.

Pérez, A.G., Sanz, C., Olías, R., Ríos, J.J. and Olías, J.M. (1996) Evolution of strawberry alcohol acyltransferase activity during fruit development and storage. *Journal of Agricultural Food Chemistry* 44, 3286–3290.

Pérez, A.G., Sanz, C. Olías, R., Ríos, J.J. and Olías, J.M. (1997) Aroma quality evaluation of strawberry cultivars in southern Spain. *Acta Horticulturae* 439, 337–340.

Peries, O.S. (1992) Studies on strawberry mildew caused by *Sphaenotheca macularis* (Wallr. Ex Fries) Jaczewski. II. Host-parasite relationships on foliage of strawberry varieties. *Annals of Applied Biology* 50, 225–233.

Perkins-Veazie, P. (1995) Growth and ripening of strawberry fruit. *Horticultural Reviews* 107, 265–295.

Perkins-Veazie, P. (1996) Strawberry physiology and maintenance of quality. In: Pritts, M.P., Chandler, C.K. and Crocker, T.E. (eds) *Proceedings of the IV North American Strawberry Conference*, Orlando, Florida.

Perkins-Veazie, P. and Collins, J.K. (1995) Strawberry fruit quality and its maintenance in postharvest environments. *Advances in Strawberry Research* 14, 1–8.

Perry, S. and Ramsdell, D. (1983) *Strawberry Diseases in Michigan*. Extension Bulletin E-1728, Cooperative Extension Service, Michigan State University, East Lansing.

Pesis, E. and Avissar, I. (1990) Effect of postharvest application of acetaldehyde vapor on strawberry decay, taste and certain volatiles. *Journal of Science, Food and Agriculture* 52, 377–385.

Peterson, R.M. (1953) Breeding behavior of the strawberry with respect to time of blooming, time of ripening, and rate of fruit development. PhD dissertation, University of Minnesota.

Pettitt, T.R. and Pegg, G.F. (1994) Sources of crown rot (*Phytophthora cactorum*) infection in strawberry and the effect of cold storage on susceptibility to the disease. *Annals of Applied Biology* 125, 279–292.

Picon, A., Martinez-Javega, J.M., Cuquerella, J., Del Rio, M.A. and Navarro, P. (1993) Effects of precooling, packaging film, modified atmosphere and ethylene absorber on the quality of refrigerated Chandler and Douglas strawberries. *Food Chemistry* 48, 189–193.

Pickel, C, Zalom, F.G., Walsh, D.B. and Welch, N.C. (1995) Vacuums provide limited *Lygus* control in strawberries. *California Agriculture* 49, 19–22.

Piringer, A.A. and Scott, D.H. (1964) Interrelation of photoperiod, chilling, and flower cluster and runner production by strawberries. *Proceedings of the American Society for Horticultural Science* 84, 295–301.

Poling, E. B. (1991) The annual hill planting system for southeastern North Carolina. In: Dale, A. and Luby, J.J. (eds) *The Strawberry into the 21st Century*. Timber Press, Portland, Oregon, pp. 258–264.

Poling, E.B. (1994) Strawberry plasticulture in North Carolina. Part II. Preplant, planting and postplant considerations for growing 'Chandler' strawberry on black plastic mulch. *Journal of Small Fruit and Viticulture* 2, 53–79.

Poling, E.B. and Parker, K. (1990) Plug production of strawberry transplants. *Advances in Strawberry Production* 9, 37–39.

Ponappa, T., Scheerens, J.C. and Miller, A.R. (1993) Vacuum infiltration of polyamines increases firmness of strawberry slices under various storage conditions. *Journal of Food Science* 58, 361–364.

Pooler, M.R., Ritchie, D.F. and Hartung, J.S. (1996) Genetic relationships among strains of *Xanthomonas fragariae* based on random amplified polymorphic DNA PCR, repetitive extragenic palindromic PCR, and enterobacterial repetitive intergenic consensus PCR data and generation of multiplexed PCR primers useful for the identification of this phytopathogen. *Applied Environmental Microbiology* 62, 3121–3127.

Popenoe, J. and Swartz, H.J. (1985) Yield component comparison of strawberry plants grown in various cultural systems. *Advances in Strawberry Production* 4, 10–14.

Popenoe, W. (1921) The frutilla, or Chilean strawberry. *Journal of Heredity* 12, 457–466.

Popenoe, W. (1926) Round about Bogotá: a hunt for new fruits and plants among the mountain forests of Colombia's unique capital. *National Geographic* XLIX, 127–160.

Popescu, A.N., Isac, V.S., Coman, M.S. and Radulescu, M.S. (1997) Somaclonal variation in plants regenerated by organogenesis from callus culture of strawberry (*Fragaria ananassa*). *Acta Horticulturae* 439, 89–96.

Popova, I.V., Konstantinova, A.E., Zekalashivili, A.U. and Zhananov, B.K. (1985) Features of breeding strawberries for resistance to berry molds. *Soviet Agricultural Science* 3, 29–33.

Porebski, S. and Catling, P.M. (1997) Intraspecific classification of *Fragaria chiloensis* for germplasm protection and utilization. *American Journal of Botany* 84, 223.

Posnette, A.F. (1953) Heat inactivation of strawberry viruses. *Nature (London)* 169, 837–838.

Posnette, A.F. and Chiykowski, L.N. (1987) Strawberry green petal and similar diseases. In: Converse, R.H. (ed.)*Virus Diseases of Small Fruits*. US Department of Agriculture Handbook 631, pp. 34 – 38.

Potter, J.W. (1995) New approaches in nematode management. In: Pritts, M.P., Chandler, C.K. and Crocker, T.E. (eds) *Proceedings of the IV North American Strawberry Conference*, University of Florida, Orlando, pp. 72–78.

Potter, J.W. and Dale, A. (1994) Wild and cultivated strawberries can tolerate or resist root-lesion nematode. *HortScience* 29, 1074–1077.

Potter, D., Luby, J. and Harrison, R.E. (1997) Phylogenetic relationships in *Fragaria* L. (Rosaceae) inferred from non-coding nuclear and chloroplast DNA sequences. *American Journal of Botany* 84, 223.

Powers, L. (1944) Meiotic studies of crosses between *Fragaria ovalis* and *F. ananassa*. *Journal of Agricultural Research* 69, 435–448.

Powers, L. (1945) Strawberry breeding studies involving crosses between the cultivated varieties (*Fragaria* × *ananassa*) and the native Rocky Mountain strawberry (*F. ovalis*). *Journal of Agricultural Research* 70, 95–122.

Powers, L. (1954) Inheritance of period of blooming in progenies of strawberries. *Proceedings of the American Society for Horticultural Science* 64, 293–298.

Pratt, H.F., Sistrunk, W.A. and Morris, J.R. (1986) Factors influencing the quality of canned strawberry filling following storage. *Journal of Food Processing and Preserves* 10, 215–226.

Prasad, K. and Stadelbacker, G.J. (1974) Effect of acetaldehyde vapor on postharvest decay and market quality of fresh strawberries. *Phytopathology* 64, 942–948.

Pringle, G.J. and Shaw, D.V. (1998) Predicted and realized response of strawberry production traits to selection in differing environments and production systems. *Journal of the American Society for Horticultural Science* 123, 61–68.

Price, J.F. and Nguyen, F.Q. (1997) Fitness of commercially produced, shipped and stored *Phytoseiulus persimilis* (Acari: Phytoseiidae) predators for dispersing in Florida strawberry. *Acta Horticulturae* 439, 913–916.

Pritts, M.P. (1988) Effect of time of defoliation on a strawberry planting damaged by root weevils. *Advances in Strawberry Production* 7, 45–46.

Pritts, M.P. (1998) Strawberry nutrition and nutrient deficiencies. In: Mass, J.L. (ed.) *Compendium of Strawberry Diseases*, APS Press, St Paul, Minnesota, 98 pp.

Pritts, M.P. and Handley, D. (1998) *Strawberry Production Guide for the Northeast, Midwest and Eastern Canada*. NRAES–88, Northeast Regional Agricultural Engineering Service, Ithaca, New York.

Pritts, M.P. and Kelly, M.J. (1993) Alternative weed management strategies for strawberries. *Acta Horticulturae* 348, 321–327.

Pritts, M.P. and Worden, K.A. (1988) Effects of duration of flower and runner removal on productivity of three photoperiodic types of strawberries. *Journal of the American Society for Horticultural Science* 113, 185–189.

Pritts, M.P., Worden, K.A. and Bartsch, J.A. (1987) Factors influencing quality and shelf life of eastern strawberry cultivars. *Cornell University Report No. 87–1*, Ithaca, New York.

Pritts, M. P., Worden, K.A. and Sheavly, M. (1989) Rowcover material and time of application and removal affect ripening and yield of strawberries. *Journal of the American Society for Horticultural Science* 114, 531–536.

Pritts, M.P., Kelly, M.J. and Eames-Sheavly, M. (1992) Modifications of renovation practices in strawberry. *Advances in Strawberry Production* 11, 28–31.

Proebsting, E. L. Sr. (1957) The effect of soil temperature on mineral nutrition of the strawberry. *Proceedings of the American Society for Horticultural Sciences* 69, 278–281.

Pyysalo, T., Honkanen, E. and Hirvi, T. (1979) Volatiles of wild strawberries, *Fragaria vesca* L. compared to those of cultivated strawberries, *Fragaria × ananassa* cv. Senga sengana. *Journal of Agricultural Food Chemistry* 27, 19–22.

Ranwala, A.P., Suematsu, C. and Masuda, H. (1992) Soluble and wall-bound invertases in strawberry fruit. *Plant Science* 84, 59–64.

Raworth, D.A. and Strong, W.B. (1994) Development of a management protocol for the twospotted spider mite, *Tetranychus urticae* Koch (Acari: Tetranychidae) on strawberries. In: Bostanian, N.J., Wilson, L.T. and Dennehy, T.J. (eds) *Monitoring and Integrated Management of Arthropod Pests of Small Fruit Crops.* Intercept Ltd, Andover, UK, pp. 103–116.

Recupero, S, Faedi, W., Rosati, P. and Arcuti, P. (1983) Results of the southern Italy breeding program. *Acta Horticulturae* 348, 151–154.

Reddy, A.S.N., Jena, P.K., Mukherjee, S.K. and Poovaiah, B.W. (1990) Molecular cloning of cDNAs for auxin-induced mRNAs and developmental expression of the auxin-inducible genes. *Plant Molecular Biology* 14, 643–653.

Reddy, A.S.N. and Poovaiah, B.W. (1987) Accumulation of a glycine rich protein in auxin-deprived strawberry fruits. *Biochemical and Biophysical Research Communications* 147, 885–891.

Reddy, A.S.N. and Poovaiah, B.W. (1990) Molecular cloning and sequencing of a cDNA for an auxin-repressed mRNA: correlation between fruit growth and repression of the auxin-regulated gene. *Plant Molecular Biology* 14, 127–136.

Renquist, A.R. and Hughes, H.G. (1985) Strawberry cultivar evaluation in Colorado: 1982–1984. *Advances in Strawberry Production* 4, 53–55.

Renquist, A.R., Breen, P.J. and Martin, L.W. (1982a) Influences of water status and temperature on leaf elongation in strawberry. *Scientia Horticulturae* 18, 77–85.

Renquist, A.R., Breen, P.J. and Martin, L.W. (1982b) Vegetative growth response of 'Olympus' strawberry to polyethylene mulch and drip irrigation regimes. *Journal of the American Society for Horticultural Science* 107, 369–372.

Reynolds, K.M., Ellis, M.A. and Madden, L.V. (1987) Progress in development of a strawberry leather rot forecasting system. *Advances in Strawberry Production* 6, 18–22.

Reynolds, K.M., Madden, L.V., Reichard, D.L. and Ellis, M.A. (1989) Splash dispersal of *Phytophthora cactorum* from infected strawberry fruit by simulated canopy drip. *Phytopathology* 79, 425–432.

Richardson, C.W. (1914) A preliminary note on the genetics of *Fragaria. Journal of Genetics* 3, 171–177.

Ricketson, C. L. (1968) 'Solid-bed' plantings – a promising method of growing strawberries. *Horticulture Research Institute of Ontario Annual Report for 1987*, 15–22.

Risser, G. (1997) Effect of low temperatures on pollen production and germination in strawberry. *Acta Horticulturae* 439, 651–658.

Roberts, A. N. and Kenworthy, A.L. (1956) Growth and composition of the strawberry plant in relation to root temperature and intensity of nutrition. *Proceedings of the American Society for Horticultural Science* 68, 157–168.

Rodriquez-Kabana, R., Morgan-Jones, G. and Chet, I. (1987) Biological control of nematodes: soil amendments and microbial antagonists. *Plant and Soil* 100, 237–247.

Rogers, W.S. and Modlibowska, I. (1951) Strawberry cultivation studies. III. Spaced and matted systems. *Journal of Horticultural Science* 26, 47–59.

Rosati, P. (1991) The strawberry in Europe. In: Dale, A. and Luby, J.J. (eds) *The Strawberry into the 21st Century*. Timber Press, Portland, Oregon, pp. 27–35.

Rosati, P. (1993) Recent trends in strawberry production and research: an overview. *Acta Horticulturae* 348, 23–44.

Rose, J.B. (1992) A possible cause of June yellows – a degenerative, non-infectious condition. *Plant Pathology* 41, 379–383.

Rosen, J.C., and Kader, A.A. (1989) Postharvest physiology and quality maintenance of sliced pear and strawberry fruits. *Journal of Food Chemistry* 54, 656–659.

Rotthoff, W. (1980) Challenging practices, systems and thoughts for the '80's'. In: Childers, N.F. (ed.) *The Strawberry: Cultivars to Marketing*. Horticultural Publications, Gainesville, Florida, pp. 77–81.

Roudeillac, P. (1983) Objectives and methods of the strawberry breeding program sponsored by the CIREF. *Acta Horticulturae* 348, 117–123.

Roudeillac, P. and Markocic, M. (1997) Improved flavour and soil disease tolerance: two major aims of the breeding program at CIREF for new advanced selections combining good commercial potential. *Acta Horticulturae* 439, 139–144.

Ruef, J.V. and Richey, H.W. (1925) A study of flower bud formation in the Dunlap strawberry. *Proceedings of the American Society for Horticultural Science* 22, 252–260.

Ruff, J.H. and Holmes, R.G. (1976) Factors affecting selectivity in the air-suspension, stem-vibration strawberry harvestor concept. *Transactions of the American Society of Agricultural Engineers* 19, 21–26.

Rwabahizi, S. and Wrolstad, R.E. (1988) Effects of mold contamination and ultra-filtration on the color stability of strawberry juice and concentrate. *Journal of Food Science* 3, 857–861.

Sacks, E.J. and Shaw, D.V. (1993) Color change in fresh strawberry fruit of seven genotypes stored at 0°C. *HortScience* 28, 209–210.

Sakin, M., Hancock, J.F. and Luby, J.J. (1997) Identifying new sources of genes that determine cyclic flowering in Rocky Mountain populations of *Fragaria virginiana* ssp. *glauca* Staudt. *Journal of the American Society for Horticultural Science* 122, 205–210.

Saks, Y., Copel, A. and Barkai-Golan, R. (1996) Improvement of harvested strawberry quality by illumination: color and *Botrytis* infection. *Postharvest Biology and Technology* 8, 19–27.

Sampson, A.C. (1994) Control of *Otiorhynchus sulcatus* in soft fruits using drench treatments of *Steinernema carpocapsae*. In: *Proceedings Brighton Crop Protection Conference, Pest and Diseases 1994* pp. 601–608.

Sanchez Egiualde, D. (1997) New strawberry cultivars:selection and development program of Planta de Navarra. S.A. *Acta Horticulturae* 439, 227–232.

Sangiacomo, M.A. and Sullivan, J.A. (1994). Introgression of wild species into the cultivated strawberry using synthetic octoploids. *Theoretical and Applied Genetics* 88, 349–354.

Sansavini, S., Rosati, P., Gaggioli, D. and Toschi, M.F. (1990) Inheritance and stability of somaclonal variations in micropropagated strawberry. *Acta Horticulturae* 280, 375–384.

Sanz, C., Pérez, A.G. and Richardson, D.G. (1994) Simultaneous HPLC determination of 2,5-dimethyl–4-hydroxy–3(2H)-furanon and related flavor components in strawberries. *Journal of Food Science* 59, 139–141.

Sato, T., Hanaoka, T., Takai, T. and Henmi, S. (1965) Studies on the breeding of strawberries adapted to the northern part of Japan, 2: Expression of some characters in S_1 and F_1 populations. *Bulletin of the Horticultural Research Station, Series C3, Morioka.*

Sauer, J.D. (1993) *Historical Geography of Crop Plants.* CRC Press, Boca Raton, Florida.

Save, R., Penuelas, J., Marfa, O. and Serrano, L. (1993) Changes in leaf osmotic and elastic properties and canopy structure of strawberries under mild water stress. *HortScience* 28, 925–927.

Saxena, G.K. and Locacio, S.J. (1968) Fruit quality of fresh strawberries as influenced by nitrogen and potassium nutrition. *Proceedings of the American Society for Horticultural Sciences* 92, 354–362.

Scaeffer, G. W., Damiano, C., Scott, D.H., McGrew, J.R., Krul, W.R. and Zimmerman, R.H. (1980) Transcription of panel discussion on the genetic stability of tissue culture plants. In: *Proceedings of a Conference on Nursery Production of Fruit Plants Through Tissue Culture – Applications and Feasibility.* USDA-SEA, ARR-NE–11, pp. 64–79.

Schaefers, G.A. (1980) Yield effects of tarnished plant bug feeding on June-bearing strawberry varieties in New York State. *Journal of Economic Entomology* 59, 721–725.

Schaefers, G.A. (1990) Pest management systems for strawberry insects. In: Pimentel, E. (ed.) *Handbook of Pest Management in Agriculture.* CRC Press, Boca Raton, Florida, pp. 377–393.

Schaffer, B., Barden, J.A. and Williams, J.M. (1985) Partioning of [14C]-photosynthate in fruiting and deblossomed day-neutral strawberry plants. *Hort-Science* 20, 911–913.

Schaffer, B., Barden, J.A. and Williams, J.M. (1986a) Net photosynthesis, dark respiration, stomatal conductance, specific leaf weight, and chlorophyll content of strawberry plants as influenced by fruiting. *Journal of the American Society for Horticultural Science* 111, 82–86.

Schaffer, B., Barden, J.A. and Williams, J.M. (1986b) Whole plant photosynthesis and dry matter partitioning in fruiting and deblossomed day-neutral strawberry plants. *Journal of the American Society for Horticultural Science* 113, 430–433.

Scheerens, J.C. and Stetson, J.F. (1996) Why do strawberries taste and smell so good? In: Pritts, M.P., Chandler, C.K. and Crocker, T.E. (eds) *Proceedings of the IV. North American Strawberry Conference,* Orlando, Florida.

Scheel, D.C. (1982) Ribbon row or close spaced strawberry plantings: a summary of growers results. *Advances in Strawberry Production* 1, 27–28.

Schiemann, E. (1937) Artkreuzungen bei Fragaria. II. *Zeitschr. Indukt. Abstamm. U. Vererbungsl.* 73, 375–390.

Schoen, C.D. and Leone, G. (1995) Towards molecular detection methods for aphid borne strawberry viruses. *Acta Horticulturae* 385, 55–63.

Schreier, P. (1980) Quantitative composition of volatile constituents in cultivated strawberries, *Fragaria × ananassa* cv. Senga sengana, Senga litessa and Senga gourmella. *Journal of Food Science* 31, 487–494.

Schuster, D.J., Price, J.F., Martin, F.G., Howard, C.M. and Albergts, E.E. (1980)

Tolerance of strawberry cultivars to two-spotted spider mites in Florida. *Journal of Economic Entomology* 73, 52–54.

Scoggan, H. J. (1978) *The Flora of Canada*. National Museum of Canada, Ottawa.

Scott, D. H. (1951) Cytological studies on polyploids derived from *Fragaria vesca* and cultivated strawberries. *Genetics* 36, 311–331.

Scott, D.H. (1959) Size, firmness, and time of ripening of fruit of seedlings of *Fragaria virginiana* Duch. crossed with cultivated strawberry varieties. *Proceedings of the American Society for Horticultural Science* 74, 388–393.

Scott, D.H., Darrow, G.M., Jeffers, W.F. and Ink, D.K. (1950) Further studies on breeding strawberries for resistance to red stele disease. *Transactions of the Peninsula Horticulture Society* 40, 1–9.

Scott, D.H., Draper, A.D. and Greeley, L.W. (1972) Interspecific hybridization in octoploid strawberries. *HortScience* 7, 382–384.

Scott, D.H., Knight, R.J. and Waldo, G.F. (1962a) Blossom sterility of strawberry seedlings. *Journal of Heredity* 53, 187–191.

Scott, D.H., Lawrence, F.J. and Draper, A.D. (1979) Strawberry varieties in the United States. *USDA Farmers' Bulletin no. 1043*.

Scott, D.H., Stembridge, G.E. and Converse, R.H. (1962b) Breeding studies with *Fragaria* for resistance to red stele root rot (*Phytophthora fragariae* Hickman) in the United States of America. *Proceedings of the International Horticulture Congress* 16, 92–98.

Sebastiampilla, A.R. and Jones, J.K. (1976) Improved techniques for the induction and isolation of polyploids in the genus *Fragaria*. *Euphytica* 25, 725–734.

Seemuller, E. (1977) Resistenzverhalten von Erdbeersorten gegen den Erreger der Rhizomfaule, *Phytophthora cactorum*. *Nachrichtenbl. Deutsche Pflanzenschutzd. (Braunschweig)* 29, 124–126.

Senanayake, Y.D.A. and Bringhurst, R.S. (1967) Origin of *Fragaria* polyploids. I. Cytological evidence. *American Journal of Botany* 54, 221–228.

Serrano, L., Carbonell, X., Save, R., Marfa, O. and Penuelas, J. (1992) Effects of irrigation regimes on the yield and water use of strawberry. *Irrigation Science* 13, 45–48.

Shamaila, M., Baumann, T.E., Eaton, G.W., Powrie, W.D. and Skura, B.J. (1992) Quality attributes of strawberry cultivars grown in British Columbia. *Journal of Food Science* 57, 696–699.

Shanks, C.H., Jr. and Barritt, B.H. (1974) *Fragaria chiloensis* clones resistant to the strawberry aphid. *HortScience* 9, 202–203.

Shanks, C.H., Jr and Barritt, B.H. (1980) Twospotted spider mite resistance of Washington strawberries. *Journal of Economic Entomology* 73, 419–423.

Shanks, C.H., Jr and Barritt, B.H. (1984) Resistance of *Fragaria chiloensis* clones to the twospotted spider mite. *HortScience* 19, 640–641.

Shanks, C.H., Jr and Doss, R.P. (1986) Black vine weevil (Coleoptera: Curculionidae) feeding and oviposition on leaves of weevil-resistant and -susceptible strawberry clones presented in various quantities. *Environmental Entomology* 15, 1074–1077.

Shanks, C.H., Jr and Moore, P.P. (1995) Resistance to twospotted spider mite and the strawberry aphid in *Fragaria chiloensis*, *F. virginiana* and *F. × ananassa* clones. *HortScience* 30, 596–599.

Shanks, C.H., Jr, Chase, D.L. and Chamberlain, J.D. (1984) Resistance of clones of wild strawberry, *Fragaria chiloensis*, to adult *Otiorhynchus sulcatus* and *O. ovatus* (Coleoptera: Curculionidae). *Environmental Entomology* 13, 1042–1045.

Shaw, D.V. (1988) Genotypic variation and genotypic correlations for sugars and organic acids of strawberries. *Journal of the American Society for Horticultural Science* 113, 770–774.

Shaw, D.V. (1989) Variation among heritability estimates for strawberries obtained by offspring–parent regressions with relatives raised in separate environments. *Euphytica* 44, 157–162.

Shaw, D.V. (1991) Recent advances in the genetics of strawberries. In: Dale, A. and Luby, J.J. (eds) *The Strawberry into the 21st Century*. Timber Press, Portland, Oregon, pp. 76–83.

Shaw, D.V. (1995) Comparison of ancestral and current-generation inbreeding in an experimental strawberry breeding population. *Theoretical and Applied Genetics* 90, 237–241.

Shaw, D.V. and Larson, K.D. (1996) Relative performance of strawberry cultivars from California and other North American sources in fumigated and non-fumigated soils. *Journal of the American Society for Horticultural Science* 121, 764–767.

Shaw, D.V. and Sacks, E.J. (1995) Response in genotypic and breeding value to a single generation of divergent selection for fresh fruit color in strawberry. *Journal of the American Society for Horticultural Science* 120, 270–273.

Shaw, D.V., Bringhurst, R.S. and Voth, V. (1987) Genetic variation for quality traits in an advanced-cycle breeding population of strawberries. *Journal of the American Society for Horticultural Science* 112, 699–702.

Shaw, D.V., Bringhurst, R.S. and Voth, V. (1988) Quantitative genetic variation for resistance to leaf spot (*Ramularia tulasnei*) in California strawberries. *Journal of the American Society of Horticultural Science* 113, 451–456.

Shaw, D.V., Bringhurst, R.S. and Voth, V. (1989) Genetic parameters estimated for an advanced-cycle strawberry breeding population at two locations. *Journal of the American Society for Horticultural Science* 114, 823–827.

Shaw, D.V., Gubler, W.D., Larson, K.D. and Hansen, J. (1996) Genetic variation for field resistance to *Verticillium dahliae* evaluated using genotypes and segregating progenies of California strawberries. *Journal of the American Society for Horticultural Science* 121, 625–628.

Sherman, W.B., Janick, J. and Erickson, H.T. (1966) Inheritance of fruit size and maturity in strawberry. *Proceedings of the American Society for Horticultural Science* 89, 309–317.

Shoemaker, N.P., Swartz, H.J. and Galletta, G.J. (1985) Cultivar dependent variation in pathogen resistance due to tissue culture-propagation of strawberries. *HortScience* 20, 253–254.

Silva, T. and Jones, J.K. (1996) In vitro propagation of *Fragaria vesca* × *Potentilla fruticosa* hybrids. *Plant Cell Tissue Organ Culture* 46, 51–58.

Simovic, N., Wolyn, D. and Jelenkovic, G. (1989) Molecular cloning, restriction and sequence analysis of 18s rRNA gene in octoploid strawberries. In: Dale, A. and Luby, J.J. (eds) *The Strawberry into the 21st Century*. Timber Press, Portland, Oregon, pp. 126–128.

Simpson, D.W. (1988) The inheritance of mildew resistance in everbearing and day-neutral strawberry seedlings. *Journal of Horticulture* 62, 329–334.

Simpson, D.W. (1991) Strawberry breeding in the United Kingdom. In: Dale, A. and

Luby, J.J. (eds) *The Strawberry into the 21st Century.* Timber Press, Portland, Oregon, pp. 50–52.

Simpson, D.W. (1993) The performance of North American day-neutral cultivars and the use of this germplasm for breeding in the United Kingdom. *Acta Horticulturae* 348, 124–130.

Simpson, D.W. and Sharp, R.S (1988) The inheritance of fruit yield and stolon production in everbearing strawberries. *Euphytica* 38:65–74.

Simpson, D.W., Bell, J.A. and Grabham, K.J. (1997a) Progress in breeding strawberries for an extended season in the United Kingdom. *Acta Horticulturae* 439, 133–138.

Simpson, D.W., Bell, J.A. and Grabham, K.J. (1997b) New strawberry cultivars from Horticulture Research International, East Malling. *Acta Horticulturae* 439, 245–250.

Simpson, D.W., Easterbrook, M.A., Bell, J.A. and Greenway, C. (1997c) Resistance to *Anthonomus rubi* in the cultivated strawberry. *Acta Horticulturae* 439, 211–215.

Simpson, D.W., Winterbottom, C.Q., Bell, J.A. and Maltoni, M.L. (1994) Resistance to a single UK isolate of *Colletotrichum acutatum* in strawberry germplasm from Northern Europe. *Euphytica* 77, 161–164.

Sistrunk, W.A. and Cash, J.N. (1968) Stabilizing the color of frozen strawberries. *Arkansas Farm Research* 17, 2.

Sistrunk, W.A. and Moore, J.N. (1980) Evaluating strawberry selections for mechanization and high quality. In: Martin, L.W. and Morris, J.R. (eds) *Strawberry Mechanization.* Oregon State University Agriculture Station Bulletin 645, Corvallis, pp. 133–170.

Sistrunk, W.A. and Morris, J.R. (1978) Storage stability of strawberry products manufactured from mechanically harvested strawberries. *Journal of the American Society of Horticultural Science* 103, 616–620.

Sistrunk, W.A. and Morris, J.R. (1980) Quality and utilization of machine harvested strawberries. In: Childers, N.F. (ed.) *The Strawberry: Cultivars to Marketing.* Horticultural Publications, Gainesville, Florida, pp. 373–384.

Sistrunk, W.A., Morris, J.R. and Gascoigne, H.L. (1982) Effect of storage temperature and treatment on quality changes in strawberry spread. *Arkansas Farm Research* 32, 7.

Sistrunk, W.A., Nunek, J. and Morris, J.R. (1980) Effects of mechanization on product quality and utilization of strawberries. Agricultural Experiment Station Bull. 645, Oregon State University, Corvallis, pp. 150–160.

Sjulin, T.M. and Dale, A. (1987) Genetic diversity of North American strawberry cultivars. *Journal of the American Society for Horticultural Science* 112, 375–385.

Sjulin, T.M., Robbins, J. and Barritt, B.H. (1986) Selection for virus tolerance in strawberry. *Journal of the American Society for Horticultural Science* 111, 458–464.

Skrede, G., Wrolstad, R.E., Lea, P. and Enersen, G. (1992) Color stability of strawberry and black currant syrups. *Journal of Food Science* 57, 172–177.

Slate, G.L. (1943) A second report on the best parents in strawberry breeding. *Proceedings of the American Society for Horticultural Sciences* 43, 175–179.

Smagula, J.M. and Bramlage, W.J. (1977) Acetaldehyde accumulation: is it a cause of physiological deterioration of fruits? *HortScience* 12, 200–203.

Smart, G.C. and Nguyen, K.B. (1991) Sting and awl nematodes: *Belonolaimus* spp. and *Dolichodorus* spp. In: Nickle, W.R. (ed.) *Manual of Agricultural Nematology.* Marcel Dekker, New York, pp. 727–667.

Smeda, R.J. and Putnam, A.R. (1988) Cover crop suppression of weeds and influence on strawberry yields. *HortScience* 23, 132–134.

Smeets, L. (1979) Effect of temperature and day-length on flower initiation and runner formation in two everbearing strawberry cultivars. *Scientia Horticulturae* 12, 19–26.

Smith, B.J. and Black, L.L. (1987) Resistance of strawberry plants to *Colletotrichum fragariae* affected by environmental conditions. *Plant Disease* 71, 834–837.

Smith, B.J. and Black, L.L. (1990) Morphological, cultural, and pathogenic variation among *Colletotrichum* species isolated from strawberry. *Plant Disease* 74, 69–76.

Smith, B.J. and Gupton, C.L. (1993) Strawberry parent clones US70, US159, US292, and US438 resistant to anthracnose crown rot. *HortScience* 28, 1055–1056.

Smith, R.B. (1986) Bulk storage of mechanically harvested strawberries for processing. *HortScience* 21, 478–480.

Smith, R.B. (1992) Controlled atmosphere storage of 'Redcoat' strawberry fruit. *Journal of the American Society for Horticultural Science* 117, 260–264.

Smith, R.B. and Skog, L.J. (1992) Postharvest carbon dioxide treatment enhances firmness of several cultivars of strawberry. *HortScience* 27, 420–421.

Smith, W.L. and Heinze, P.H. (1958) Effect of color development at harvest on quality of post-harvest ripened strawberries. *Proceedings of the American Society for Horticultural Science* 72, 207–211.

Smith, W., Christiamson, A. and Robe, K. (1974) Crust freezing in liquid freon doubles capacity – 20, 000 lb/hr. *Food Processing* 35, 17.

Smith, W.L., Moline, H.E. and Johnson, K.S. (1979) Studies with *Mucor* species causing postharvest decay of fresh produce. *Phytopathology* 69, 865–869.

Sohati, P.H., Boivin, G. and Stewart, R.K. (1992) Parasitism of *Lygus lineolaris* eggs on *Coronilla varia*, *Solanum tuberosum*, and three host weeds in southeastern Quebec. *Entomophaga* 37, 515–523.

Sommer, N.F., Buckley, P.M., Fortlage, R.J., Coon, D.A., Maxie, E.C. and Mitchell, F.G. (1968) Heat sensitization for control of grey mold of strawberry fruits by gamma irradiation. *Radiation Biology* 8, 441–448.

Sosa-Alvarez, M., Madden, L.V. and Ellis, M.A. (1995) Effects of temperature and wetness duration on sporulation of *Botrytis cinerea* on strawberry leaf residues. *Plant Disease* 79, 609–615.

Spangelo, L.P.S., Hsu, C.S., Fejer, S.O. and Watkins, R. (1971a) Inbred line × tester analysis and the potential of inbreeding in strawberry breeding. *Canadian Journal of Genetics and Cytology* 13, 460–469.

Spangelo, L.P.S., Hsu, C.S., Fejer, S.O., Bedard, P.R. and Rouselle, G.L. (1971b) Heritability and genetic variance components for 20 fruit and plant characters in the cultivated strawberry. *Canadian Journal of Genetics and Cytology* 13, 443–456.

Spayd, S.E. and Morris, R.S. (1981) Physical and chemical characteristics of puree from once-over harvested strawberries. *Journal of the American Society for Horticultural Science* 106, 101–105.

Spencer, D.M. (1978) *The Powdery Mildew*. Academic Press, London.

Sruamsiri, P. and Lenz, F. (1986) Photosynthesis and stomatal behavior of strawberries. V. Effect of water deficiency. *Gartenbau.* 51, 84–92.

Stadelbacker, G.J. (1963) Why so much variation in strawberry fertilizer recommendations and practices. In: Smith, C.R. and Childers, N.F. (eds) *The Strawberry*. Rutgers State University, New Brunswick, New Jersey.

Stahler, M.M. (1990) Evaluation of Minnesota and Wisconsin populations of *Fragaria virginiana* for horticultural and morphological traits. PhD Thesis. University of Minnesota, St Paul.

Stahler, M.M., Ascher, P.D., Luby, J.J. and Roelfs, A.P. (1995) Sexual composition of populations of *Fragaria virginiana* (Rosaceae) collected from Minnesota and western Wisconsin. *Canadian Journal of Botany* 73, 1457–1463.

Stahler, M.M., Luby, J.J. and Ascher, P.D. (1990) Comparative yield of female and hermaphroditic *Fragaria virginiana* germplasm collected in Minnesota and Wisconsin. In: Dale, A. and Luby, J.J. (eds) *The Strawberry into the 21st Century*. Timber Press, Portland, Oregon.

Stanisavljević, M., Srećković, M. and Mitrović, M. (1997) Field performance of some foreign strawberry cultivars grown in Yugoslavia. *Acta Horticulturae* 439, 291–295.

Staudt, G. (1952) Cytogenetische Untersuchungen an *Fragaria orientalis* Los und ihre Bedeutung für Artbildung und Geschlechtsdifferenzierung in der Gattung *Fragaria*. *Vererbungslehre* 84, 361–416.

Staudt, G. (1959) Cytotaxonomy and phylogentic relationship in the genus *Fragaria*. *Proceedings of the 9th International Congress of Botany*, p. 377.

Staudt, G. (1961) The origin of the large-fruited garden strawberry, *Fragaria × ananassa* Duch. (In German). *Der Zuchter* 31, 5–9.

Staudt, G. (1962) Taxonomic studies in the genus *Fragaria*. Typification of *Fragaria* species known at the time of Linnaeus. *Canadian Journal of Botany* 40, 869–886.

Staudt, G. (1967a) The genetics and evolution of heterosis in the genus *Fragaria* 1. Research with *Fragaria orientalis* (in German). *Zeitschrift für Planzenzüchtung* 58, 245–277.

Staudt, G. (1967b) The genetics and evolution of heterosis in the genus *Fragaria* 2. Species hybridization of *F. vesca × F. orientalis* and *F. viridis × F. orientalis* (in German). *Zeitschrift für Planzenzüchtung* 58, 309–322.

Staudt, G. (1968) The genetics and evolution of heterosis in the genus *Fragaria* 3. Research with hexaploid and octoploid kinds (in German). *Zeitschrift für Planzenzüchtung* 59, 83–102.

Staudt, G. (1973) *Fragaria iturupensis*, eine neue Erdbeerart aus Ostasien. *Willenowia* 7, 101–104.

Staudt, G. (1984) Der cytologische Nachweis von doppelter Restitution bei *Fragaria*. *Plant Systematics and Evolution* 146, 171–179.

Staudt, G. (1989) The species of *Fragaria*, the taxonomy and geographical distribution. *Acta Horticulturae* 265, 23–33.

Staudt, G. (1997) Reconstitution of *Fragaria × ananassa*; the effect of *Fragaria virginiana* cytoplasm. *Acta Horticulturae* 439, 55–62.

Staudt, G. (1999) *Systematics and Geographic Distribution of the American Strawberry Species: Taxonomic Studies in the Genus* Fragaria (Rosaceae: Potentilleae). University of California Press, Berkeley.

Stembridge, G.E. and Scott, D.H. (1959) Inheritance of resistance of strawberry to the common race of the red stele root rot fungus. *Plant Disease Reporter* 43, 1091–1094.

Stenger, D.C., Mullin, R.H. and Morris, T.J. (1988) Isolation, molecular cloning and detection of strawberry vein banding virus DNA. *Phytopathology* 78, 154–159.

Sterk, G. and Meesters, P. (1997) IPM on strawberries in glasshouses and plastic tunnels in Belgium, new possibilities. *Acta Horticulturae* 432, 905–912.

Stephens, D. (1996) No replacement for methyl bromide. *Fruit Grower* 1, 30–31.

Stirling, G.R. (1991) *Biological Control of Plant Parasitic Nematodes.* CAB International, Wallingford, UK.

Strabbioli, G. (1988) A study of strawberry water requirements. *Acta Horticulturae* 228, 179–186.

Strand, L.L. (1994) *Integrated Pest Management for Strawberries.* University of California Statewide Integrated Pest Management Project Publication 3351, Berkeley, 142 pp.

Strang, J.G., Archbold, D.D., Hartman, J.R. and Hendrix, J.W. (1985) Strawberry cultivar yield response to soil fumigation. *Advances in Strawberry Production* 4, 36–38.

Strik, B. (1985) Flower bud initiation in strawberry cultivars. *Fruit Varieties Journal* 39, 5–9.

Strik, B. and Proctor, J. (1988a) Yield component analysis of strawberry genotypes differing in productivity. *Journal of the American Society for Horticultural Science* 113, 124–129.

Strik, B. and Proctor, J. (1988b) Growth analysis of field-grown strawberries differing in yield. I. The matted row system. *Journal of the American Society for Horticultural Science* 113, 894–899.

Strik, B. and Proctor, J. (1988c) Growth analysis of field-grown strawberries differing in yield II. The hill system. *Journal of the American Society for Horticultural Science* 113, 899–904.

Sturhan, D. and Brzeski, M.W. (1991) Stem and bulb nematodes, *Ditylenchus* spp. In: Nickle, W.R. (ed.) *Manual of Agricultural Nematology.* Marcel Dekker, New York, pp. 423–464.

Stutte, G.W. and Darnell, R.L. (1987) A non-destructive developmental index for strawberry. *HortScience* 22, 218–221.

Subramanium, M.D. and Iyer, C.P.A. (1974) Performance of temperate strawberry cultivars in the tropics. *Proceedings of the XIX International Horticultural Congress,* p. 527.

Sutton, J.C. (1990) Epidemiology and management of botrytis leaf blight of onion and grey mold of strawberry: a comparative analysis. *Canadian Journal of Plant Pathology* 12, 100–110.

Sutton, J.C. (1994) Biological control of strawberry diseases. *Advances in Strawberry Research* 13, 1–12.

Sutton, J.C. and Gibson, I.A.S. (1977) *Pezizella oenotherae. Descriptions of Pathogenic Fungi and Bacteria, no. 535.* Commonwealth Mycological Institute, Kew, UK.

Sutton, J.C. and Peng, G. (1993) Biosuppression of inoculum production by *Botrytis cinerea* in strawberry leaves. In: Fokkema, N.J., Köhl, J. and Elad, Y. (eds) *Biological Control of Foliar and Postharvest Diseases.* IOBC/SPRD Bulletin Vol. 16. Monfavet, France, pp. 47–52

Swadling, I.R. and Jefferies, P. (1996) Isolation of microbial antagonists for biological control of grey mould disease of strawberries. *Biocontrol Science and Technology* 6, 125–136.

Swartz, H. J., Galletta, G.J. and Zimmerman, R.H. (1981) Field performance and phenotypic stability of tissue culture-propagated strawberries. *Journal of the American Society for Horticultural Science* 106, 667–673.

Swartz, H.J., Popenoe, J. and Fiola, J.A. (1985) Yield component analysis of the 1984

Maryland-USDA replicated trials. *Advances in Strawberry Production* 4, 45–52.

Szczygiel, A. (1981a) Trials on susceptibility of strawberry cultivars to the needle nematode *Longidorus elongatus*. *Fruit Science Reports* 8, 127–131.

Szczygiel, A. (1981b) Trials on susceptibility of strawberry cultivars to northern root-knot nematode, *Meloidogyne hapla*. *Fruit Science Reports* 8, 115–119.

Szczygiel, A. (1981c) Trials of strawberry cultivars to the root lesion nematode *Pratylenchus penetrans*. *Fruit Science Reports* 8, 121–125.

Szczygiel, A. and Danek, J. (1984) Trials of breeding strawberry, *Fragaria grandiflora*, cultivars resistant to the northern root-knot nematode, *Meloidogyne hapla*. *Fruit Science Reports* 11, 79–85.

Takahashi, H., Takai, T. and Matsumoto, T. (1990) Susceptible strawberry cultivars to Alternaria black spot of strawberry (*Alternaria alternaria* Strawberry Pathotype) in Japan. *Journal of the Japanese Society for Horticultural Science* 59, 539–544.

Takeda, F., Janisiewicz, W.J., Roitman, J., Mahoney, N. and Abeles, F.B. (1990) Pyrrolnitrin delays postharvest rot in strawberries. *HortScience* 25, 320–322.

Tehranifar, A., Le Mière, P. and Battey, N.H. (1998) The effects of lifting date, chilling duration and forcing temperature on vegetative growth and fruit production in the June-bearing cultivar 'Elsanta'. *Journal of Horticultural Science and Bio-technology* 73, 453–460.

Tezuka, N. and Makino, T. (1991) Biological control of Fusarium wilt of strawberry by nonpathogenic *Fusarium oxysporum* isolated from strawberry. *Annals of the Phytopathological Society of Japan* 57, 506–511.

Thomas, P. (1986) Radiation preservation of foods of plant origin. Part V. Temperate fruits: pome fruits, stone fruits and berries. *CRC Critical Reviews of Food Science and Nutrition* 24, 357–400.

Timberlake, C.F. and Bridle, P. (1982) Distribution of anthocyanins in food plants. In: Markakis, P. (ed.) *Anthocyanins as Food Colors*. Academic Press, New York, pp. 137.

Tingey, W.M. (1976) Survey of crop resistance to lygus bug. In: Scott, D.R. and O'Keefe (eds) *Lygus Bug: Host Plant Interactions*. Proceedings of a Workshop, XV International Congress of Entomology.

Toyoda, H., Morimoto, M. and Ouchi, S. (1994) Immobilization of chitin-degraded microbe in alginate gel beads and its application to suppression of fungal pathogens in soil. *Bulletin of the Institute for Comprehensive Agricultural Science, Kinki University* 2, 21–28.

Trajkovski, K. (1997) Further work on species hybridization in *Fragaria* at Balsgård, Sweden. *Acta Horticulturae* 439, 67–73.

Trumble, J.T. and Morse, J.P. (1993) Economics of integrating the predaceous mite, *Phytoseiulus persimilis* (Acari: Phytoseiidae) with pesticides in strawberries. *Journal of Economic Entomology* 86, 879–885.

Trumble, J.T., Oatman, E.R. and Voth, V. (1983a) Thresholds and sampling for aphids in strawberries. *California Agriculture* 37, 11–12.

Trumble, J.T., Oatman, E.R. and Voth, V. (1983b) Development and estimation of aphid populations infesting annual winter plantings of strawberries in California. *Journal of Economic Entomology* 76, 496–501.

Udagawa, Y., Ito, T. and Gomi, K. (1989) Effects of root temperature on some physiological and ecological characteristics of strawberry plants 'Reiko' grown in nutrient solution. *Journal of the Japanese Society of Horticultural Science* 58, 627–632 (Japanese with English summary).

Ueda, Y. and Bai, J.H. (1993) Effect of short term exposure of elevated CO_2 on flesh firmness and ester production of strawberry. *Japanese Journal of Horticultural Science* 62, 457–464.

Ulrich, A., Mostafa, M.A.E. and Allen, W.W. (1980) *Strawberry Deficiency Symptoms: a Visual and Plant Analysis Guide to Fertilization.* University of California Agricultural Sciences Publication 4098.

Vainio, A. and Hokkanen, H.M.T. (1993) The potential of entomopathogenic fungi and nematodes against *Otiorhynchus ovatus* L. and *O. dubius* Strom in the field. *Journal of Applied Entomology* 115, 379–387.

Valleau, W.D. (1918) Sterility in the strawberry. *Journal of Agricultural Research* 12, 613–669.

Van Adrichem, M.C.J. and Orchard, W.R.. (1958) Verticillium wilt resistance in the progenies of *Fragaria chiloensis* from Chile. *Plant Disease Reporter* 42, 1391–1393.

Van Adrichem, M.C.J. and Bosher, J.E. (1962) *Rhizoctonia solani* Kuhn as a component of the strawberry root rot complex in British Columbia. *Canadian Plant Disease Survey* 42, 118–121.

Van Adrichem, M.C.J., and Orchard, W.R. (1958) Verticillium wilt resistance in the progenies of *Fragaria chiloensis* from Chile. *Plant Disease Reporter* 42, 1391–1393.

Van de Lindeloof, C.P.J. and Meulenbroek, E.J. (1997) Fifty years of strawberry breeding at the Centre for Plant Breeding and Reproduction Research (CPRO-DLO) in the Netherlands. *Acta Horticulturae* 439, 115–120.

Van der Scheer, H.A.T. (1973) Susceptibility of strawberry to isolates of *Phytophthora cactorum* and *Phytophthora citricola.* Meded. Fac. *Landbauwwet Rijksuniv Gent* 38, 1407–1415.

Van der Scheer, H.A.T. (1981) Gnomonia-vruchtrot: Een toenemend probleem bij de teelt van oardbeien in Nederland. *Fuitteelt* 71, 1240–1241.

Van de Vrie, M. and Price, J.F. (1994) *Manual for Biological Control of Two-spotted Spider Mite in Florida.* DOV 1994–1, University of Florida, Dover.

Van de Weg, W.E. (1997) A gene-for-gene model to explain interactions between cultivars of strawberry and races of *Phytophthora fragariae* var *fragariae.* *Theoretical and Applied Genetics* 94, 445–451.

Van de Weg, W. E., Giezen, S., Maas, J.L. and Galletta, G.J. (1993) Identification of genes for resistance to *Phytophthora fragariae* in strawberry. *Acta Horticulturae* 348, 137–138.

Van de Weg, W.E., Henken, B. and Giesen, S. (1997) Assessment of the resistance to *Phytophthora fragariae* var *fragariae* of the USA and Canadian differential series of strawberry genotypes. *Journal of Phytopathology* 145, 1–6.

Van Rijbroek, P.C.L., Meulenbroek, E.J. and Van de Lindeloof, C.P.J. (1997) Development of a screening method for resistance to *Phytophthora cactorum.* *Acta Horticulturae* 439, 181–183.

Varney, E.H., Moore, J.N. and Scott, D.H. (1959) Field resistance of various strawberry varieties and selections to *Verticillium.* *Plant Disease Reporter* 43, 567–569.

Vaughn, S.F., Spencer, G.F. and Shasha, B.S. (1993) Volatile compounds from raspberry and strawberry fruit inhibit postharvest decay fungi. *Journal of Food Science* 58, 793–796.

Veluthambi, K. and Poovaiah, B.W. (1984) Auxin-regulated polypeptide changes at different stages of strawberry fruit development. *Plant Physiology* 75, 349–353.

Vincent, C., de Oliveira, D.D. and Bélanger, A. (1994) The management of insect pollinators and pests in Quebec strawberry plantations. In: Bostanian, N.J., Wilson, L.T. and Dennehy, T.J. (eds) *Monitoring and Integrated Management of Arthropod Pests of Small Fruit Crops*. Intercept Ltd, Andover, UK, pp.177–192.

Voth, V. (1986) Efficient use of nitrogen fertilizers in California strawberry production. In: Burns, E. and Burns, B. (eds) *Proceedings 1986 Winter Conference of the North American Strawberry Growers Association*, Tarpon Springs, Florida, pp. 31–38.

Voth, V. and Bringhurst, R.S. (1990) Culture and physiological manipulation of California strawberries. *HortScience* 25, 889–892.

Voth, V., Urin, K. and Bringhurst, R.S. (1967) Effect of high nitrogen applications on yield, earliness, fruit quality and leaf composition of California strawberries. *Proceedings of the American Society for Horticultural Science* 91, 249–256.

Vrain, T.C. (1990) The effect of an organic amendment on control of *Pratylenchus penetrans* with a non-fumigant nematoside and a parasitic fungus. *Nematologica* 36, 399.

Waldo, G.F. (1939) Effects of leaf removal and crown covering on the strawberry plant. *Proceedings of the American Society for Horticultural Sciences* 37, 548–552.

Wallin, A. (1997) Somatic hybridization in *Fragaria*. *Acta Horticulturae* 439, 63–66.

Wang, D., Wergin, W.P. and Zimmerman, R.H. (1984) Somatic embryogenesis and plant regeneration from immature embryos of strawberry. *HortScience* 19, 71–72.

Wang, S.L. and Dale, A. (1990) Evaluation of strawberry cultivars for frozen sugar pack. *Advances in Strawberry Production* 9, 31–32.

Wang, Q. and Tang, H. (1994) Strawberry culture in China. *Chronica Horticulturae* 34, 5–6.

Wardlow, L.R. (1994) Biological control agents in soft fruit. In: Taylor, D. (ed.) *New Developments in the Small Fruit Industry*. Agricultural Development and Advisory Service and Horticulture Research International, Robinson College, Cambridge, pp. 31.

Warmund, M.R. (1993) Ice distribution in 'Earliglow' strawberries and crown recovery following extracellular freezing. *Journal of the American Society for Horticultural Science* 118, 644–648.

Warmund, M.R. and English, T.J. (1994) Efficacies of cryprotectants applied to 'Honeoye' strawberry plants inoculated with ice-nucleation-active bacteria. *Advances in Strawberry Research* 13, 20–25.

Warmund, M.R. and Ki, W.K. (1992) Fluctuating temperatures and root moisture content affect survival and regrowth of cold-stressed strawberry crowns. *Advances in Strawberry Research* 11, 40–46.

Wassenaar, L.M. and van der Scheer, H.A.Th. (1989) Alternaria leaf spot in strawberry. *Acta Horticulturae* 265, 575–578.

Watanabe, T., Hashimoto, K. and Sato, M. (1977) *Pythium* species associated with strawberry roots in Japan, and their role in the strawberry stunt disease. *Phytopathology* 67, 1324–1332.

Watkins, R. and Spangelo, L.P.S. (1971) Genetic components from full, half and quarter diallels for the cultivated strawberry. *Canadian Journal of Genetics and Cytology* 13, 515–521.

Watkins, R., Spangelo, L.P.S. and Bolton, A.T. (1970) Genetic variance components in cultivated strawberry. *Canadian Journal of Genetics and Cytology* 12, 52–59.

Webb, R.A., Purves, J.V., White, B.A. and Ellis, R. (1974) A critical path analysis of fruit production in strawberry. *Scientia Horticulturae* 2, 175–184.

Welch, D. P., Burkhardt, T.H., Ledebuhr, R.L. and VanEe, G.R. (1986) Computer model to evaluate the economic feasibility of mechanical harvesting and processing of solid set strawberry production in Michigan. *American Society of Agricultural Engineers Paper No. 86–1075*. American Society of Agricultural Engineers, St Joseph, Michigan.

Welsh, S. L., Atwood, N.D., Goodrich, S. and Higgins, L.C. (1987) *A Utah Flora*. Great Basin Naturalist Memoirs, No. 9. Brigham Young University, Provo, Utah.

Went, F. (1957) Chapter 9. *The Experimental Control of Plant Growth*, Vol 17, Cronica Botanica, Waltham, Massachusetts.

Wenzel, W.G. (1980) Correlation and selection index components. *Canadian Journal of Genetics and Cytology* 13, 42–50.

Wesche-Ebeling, P. and Montgomery, M.W. (1990) Strawberry polyphenoloxidase: its role in anthocyanin degradation. *Journal of Food Science* 55, 731–734.

Whitworth, J.L. (1995) The ability of some cover crops to suppress common weeds in strawberry fields. *Journal of Sustainable Agriculture* 7, 137–145.

Wilcox, W.F. (1995) New directions in research on cool soil root pathogens. In: *Proceedings of IV North American Strawberry Conference*. Orlando, Florida, pp. 34–38.

Wilcox, W.F. and Seem, R.C. (1994) Relationship between strawberry grey mold incidence, environmental variables, and fungicide applications during different periods of the fruiting season. *Phytopathology* 84, 264–270.

Wilhelm, S. (1955) Verticillium wilt of the strawberry with special reference to resistance. *Phytopathology* 45, 387–391.

Wilhelm, S. and Nelson, R.D. (1990) Pest management systems for strawberry diseases. In: Pimentel, E. (ed.) *Handbook of Pest Management in Agriculture*. CRC Press, Boca Raton, Florida, pp. 395–404.

Wilhelm, S. and Sagen, J.A. (1974) *A History of the Strawberry*. University of California Division of Agriculture Publication 4031, Berkeley.

Wilhelm, S., Storkan, R.C. and Sagen, J.E. (1961) Verticillium wilt of strawberry controlled by fumigation of soil with chloropicrin and chloropicrin-methyl bromide mixtures. *Phytopathology* 51, 744–748.

Wilkinson, J.Q., Lanahan, M.B., Conner, T.W. and Klee, H.J. (1995) Identification of mRNAs with enhanced expression in ripening strawberry fruit using polymerase chain reaction differential display. *Plant Molecular Biology* 27, 1097–1108.

Williams, H. (1955) June yellows: a genetic disease of the strawberry. *Journal of Genetics* 53, 232–243.

Williams, H. (1974) Early assessment of processing strawberries for color. *Euphytica* 26, 841–845.

Williams, R.N. and Rings, R.W. (1980) *Insect Pests of Strawberries in Ohio*. Ohio Agricultural Research Developmental Center Research Bulletin 1122, Wooster.

Williamson, S.C., Yu, H. and Davis, T.M. (1995) Shikimate dehydrogenase allozymes: inheritance and close linkage to flower color in diploid strawberry. *Journal of Heredity* 86, 74–76.

Williamson, B., Johnston, D.J., Ramanathan, V. and McNichol, R.J. (1993) A polygalacturonase inhibitor from immature raspberry fruits: a possible new approach to grey mold control. *Acta Horticulturae* 352, 601–605.

Wills, A.B. (1962) Genetical aspects of strawberry June yellows. *Heredity* 17, 361–372.

Wilson, D.J. and Rogers, W.S. (1954) Trials of burning and mowing strawberry plants after cropping. *Journal of the Horticultural Society* 29, 21–26.

Wilson, L.L., Madden, L.V. and Ellis, M.A. (1990) Influence of temperature and wetness duration on infection of immature and mature strawberry fruit by *Colletotrichum acutatum*. *Phytopathology* 80, 111–116.

Wilson, W.F., Jr. and Giamalva, M.J. (1954) Days from bloom to harvest of Louisiana strawberries. *Proceedings of the American Society for Horticultural Science* 63, 201–204.

Wing, K.B., Pritts, M.P. and Wilcox, W.F. (1995) Biotic, edaphic, and cultural factors associated with strawberry black root rot in New York. *HortScience* 30, 86–90.

Wing, K.B., Pritts, M.P. and Wilcox, W.F. (1995) Field resistance of 20 strawberry cultivars to black root rot. *Fruit Varieties Journal* 49, 94–98.

Winks, B.L. and Williams, Y.N. (1965) A wilt of strawberry caused by a new form of *Fusarium oxysporum*. *Queensland Journal of Agricultural and Animal Science* 22, 475–479.

Winter, J.D., Landon, R.H. and Alderman, W.H. (1940) Use of CO_2 to retard the development of decay in strawberry and raspberry. *Proceedings of the American Society for Horticultural Science* 37, 583–588.

Wolyn, D.J. and Jelenkovic, G. (1990) Nucleotide sequence of an alcohol dehydrogenase gene in octoploid strawberry (*Fragaria* × *ananassa* Duch.). *Plant Molecular Biology* 14, 855–857.

Worthington, J.T. (1970) *Cold Storage and Transport of Strawberry Plants*. Marketing Research Report No. 865, United States Department of Agriculture, Washington, DC.

Wright, C.J. and Sandrang, A.K. (1993) Density effects on vegetative and reproductive development in strawberry cv. Hapil. *Journal of Horticultural Science* 68, 231–236.

Wright, F. and Hughes, H.G. (1993) Hydroponic screening of strawberry for salt tolerance: Correlation with *in vitro* evaluations. *Acta Horticulturae* 348, 384–388.

Wrolstad, R.E., Putnam, T.P. and Varseveld, G.W. (1970) Color quality of frozen strawberries: effect of anthocyanin, pH, total acidity and ascorbic acid variability. *Journal of Food Science* 35, 448–452.

Wrolstad, R.E. and Shallenberger, R.S. (1981) Free sugars and sorbitol in fruits. A compilation from the literature. *Journal of the Association of Agicultural Chemists* 64, 91–103.

Wrolstad, R.E., Skrede, G., Lea, P. and Enerson, G. (1990) Influence of sugar on anthocyanin pigment stability in frozen strawberries. *Journal of Food Science* 55, 1064–1065.

Yamamoto, M., Namiki, F., Nishimura, F. and Kohmoto, K. (1985) Studies on host specific AF toxins produced by *Alternaria alternata* strawberry pathotype causing Alternaria black spot of strawberry. 3. Use of toxin for determining inheritance of disease reaction in strawberry cultivar Morioka–16. *Annals of the Phytopathology Society of Japan* 51, 530–535.

Yangi, T. and Oda, Y. (1993) Effects of photoperiod and chilling on floral formation of intermediate types between June- and everbearing strawberries. *Acta Horticulturae* 348, 339–346.

Yarnell, S.H. (1931) Genetic and cytological studies on *Fragaria*. *Genetics* 16, 422–454.

Yoshikawa, N. and Converse, R.H. (1990) Strawberry pallidosis disease: distinctive dsRNA species associated with latent infections in indicators and in diseased cultivars. *Phytopathology* 80, 543–548.

Yoshikawa, N. and Inouye, T. (1988) Strawberry viruses occuring in Japan. *Acta Horticulturae* 236, 59–67.

Yoshikawa, N., Inouye, T. and Converse, R.H. (1986) Two types of rhabdoviruses in strawberry. *Annals of the Phytopathology Society of Japan* 52, 437–444.

Yu, H. and Davis, T.M. (1995) Genetic linkage between runnering and phosphoglucoisomerase allozymes, and systematic distortion of monogenic segregation ratios in diploid strawberry. *Journal of the American Society for Horticultural Science* 120, 687–690.

Yu, L., Reitmeier, C.A., Gleason, M.L., Nonnecke, G.R., Olson, D.G. and Gladon, R.J. (1995) Quality of electron beam irradiated strawberries. *Journal of Food Science* 60, 1084–1087.

Yu, L., Reitmeier, C.A. and Love, M.H. (1996) Strawberry texture and pectin content as effected by electron beam irradiation. *Journal of Food Science* 61, 844–846.

Yu, T.S., Lu, L.T., Ku, T.C., Kuan, K.C. and Li, C.L. (1985) *Flora Reopublicae Popularis Sinicae,*. Vol. 37. Inst. Bot. Acad. Sin. Science Press, Beijing.

Yurgalevitch, C.M., Janes, H.W. and Chin, C. (1985) Somaclonal variation in *Fragaria*. *HortScience* 20, 450.

Zabetakis, I. and Holden, M.A. (1997) Strawberry flavor: analysis and biosynthesis. *Journal of the Science of Food and Agriculture* 74, 421–434.

Zacharda, M. and Hluchy, M. (1996) Biological control of two-spotted spider mite *Tetranychus urticae* on strawberries by the predatory phytoseiid mite *Typhlodromus pyri*. *Acta Horticulturae* 422, 226–230.

Zalom, F.G., Pickel, C., Walsh, D.B. and Welch, N.C. (1993) Sampling for *Lygus hesperus* (Hemiptera:Miridae) in strawberries. *Journal of Economical Entomology* 86, 1191–1195.

Zhang, B. and Archbold, D.D. (1993a) Water relations of a *Fragaria chiloensis* and *F. virginiana* selection during and after severe water stress. *Journal of the American Society for Horticultural Science* 118, 274–279.

Zhang, B. and Archbold, D.D. (1993b) Solute accumulation in leaves of a *Fragaria chiloensis* and a *F. virginiana* selection responds to water deficit stress. *Journal of the American Society of Horticultural Science* 118, 280–285.

Zheng, J. (1992) Epidemiology and biological control of strawberry leaf scorch. MSc Thesis, University of Guelph, Canada.

Zheng, J. and Sutton, J.C. (1994) Inoculum concentration, leaf age, wetness duration, and temperature in relation to infection of strawberry leaves by *Diplocarpon earlianum*. *Canadian Journal of Plant Pathology* 16, 177–186.

Zorin, M, Greer, N., Herrington, M., Hutton, D. and Ullio, L. (1997) Australia – strawberry industry and research. *Acta Horticulturae* 439, 377–383.

Zubov, A.A. and Stankevich, K.V. (1982) Combining ability of a group of strawberry varieties for quality characters of fruits (in Russian). *Soviet Genetics* 18, 732–739.

Zurawicz, E. and Stushnoff, C. (1977) Influence of nutrition on cold tolerance of 'Redcoat' strawberries. *Journal of the American Society for Horticultural Science* 102, 342–346.

Zych, C.C. (1966) Fruit maturation times of strawberry varieties. *Fruit Variety and Horticultural Digest* 20, 51–53.

Plate 1. Matted row strawberries in New York (M. Pritts).
Plate 2. Hill cultured strawberries in California (M. Pritts).
Plate 3. Earliest printed illustration of strawberry in Peter Schöffer's *Herbarius Latinus Moguntiae* (1484).
Plate 4. The 'Hovey' strawberry (Wilhelm and Sagen, 1966).

5

6

7

9

8

10

Plate 5. Nitrogen deficiency (M. Pritts).
Plate 6. Potassium deficiency (M. Pritts).
Plate 7. Boron deficiency (M. Pritts).
Plate 8. Calcium deficiency (M. Pritts).
Plate 9. Iron deficiency (M. Pritts).
Plate 10. Magnesium deficiency (M. Pritts).

11

12

13

14

Plate 11. Matted rows mulched with straw (M. Pritts).
Plate 12. Fruit growing in an English greenhouse (T. Morton).
Plate 13. Plastic tunnels in Korea (Ho-jeong Jeong).
Plate 14. Strawberries in Japanese polyhouse (K. Kawagishi).

Plate 15. Grey mould (M. Pritts).
Plate 16. Leather rot (M. Pritts).
Plate 17. Anthracnose (M. Pritts).
Plate 18. Varietal differences in susceptibility to red stele (M. Pritts).

INDEX